The Internet and Rural Development in China: The Socio-Structural Paradigm

European University Studies
Europäische Hochschulschriften
Publications Universitaires Européennes

Series XL
Communications

Reihe XL Série XL
Kommunikationswissenschaft und Publizistik
Média et Journalisme

Vol./Band 97

PETER LANG
Bern · Berlin · Bruxelles · Frankfurt am Main · New York · Oxford · Wien

Jinqiu Zhao

The Internet and Rural Development in China: The Socio-Structural Paradigm

PETER LANG
Bern · Berlin · Bruxelles · Frankfurt am Main · New York · Oxford · Wien

Bibliographic information published by Die Deutsche Bibliothek
Die Deutsche Bibliothek lists this publication in the Deutsche Nationalbibliografie; detailed bibliographic data is available on the Internet at ‹http://dnb.ddb.de›.

British Library and Library of Congress Cataloguing-in-Publication Data:
A catalogue record for this book is available from *The British Library*, Great Britain, and from *The Library of Congress*, USA

Library of Congress Cataloging-in-Publication Data

Zhao, Jinqiu.
The Internet and rural development in China : the socio-structural paradigm / Jinqiu Zhao.
p. cm. – (European university studies = Europäische Hochschulschriften = publications universitaires européennes ; Series XL)
Includes bibliographical references.
ISBN 978-3-03-911584-6 (alk. paper)
1. Internet–China. 2. Communication in rural development–China.
3. Technological innovations–Economic aspects–China.
4. Information technology–Economic aspects–China. I. Title.
HN740.Z9C69429 2008
303.48'330917340954–dc22
2008015977

ISSN 0176-3725
ISBN 978-3-03911-584-6

To Jie –
he always looks at me
rather than the books during the schooldays

Acknowledgements

Many people have contributed in one way or another to the completion of this book to whom I owe my heartfelt appreciation.

First of all, my deepest appreciation goes to Dr. Hao Xiaoming and Dr. Indrajit Banerjee. Both of them have set good examples of what researchers should be like. This is more precious to me than anything I have learned in the process of writing the book.

My thanks also go to Dr. Kavita Karan, Dr. Xu Xiaoge, Dr. Foo Tee Tuan, Dr. Lee Wai Peng, and Dr. Chan Kin Ying for their patience in answering my never-ending queries.

Table of Contents

Acknowledgements 7

Preface 13

1. Introduction 17
   - 1.1 Purpose of the Study 17
   - 1.2 Rationale and Scope of the Study 18
   - 1.3 Research Questions 22
   - 1.4 Brief Description of Methodology 22
   - 1.5 Significance of the Study 23
   - 1.6 Outline of the Book 25

2. Development Communication 27
   - 2.1 The Modernization Paradigm 27
   - 2.2 The Dependency Paradigm 37
   - 2.3 The Alternative Pathway to Development 41
   - 2.4 Summary 46

3. ICTS and Rural Development 49
   - 3.1 Technological Convergence 49
   - 3.2 The Concept of Human Development 53
   - 3.3 ICTs and Development: Optimists vs. Pessimists 56
   - 3.4 The Internet and Rural Development 59
   - 3.5 Summary 73

4. Diffusion of Innovations 75
   - 4.1 The Main Elements 76
   - 4.2 The Main Propositions 81
   - 4.3 Criticisms of the Diffusion Approach 83

4.4 A New Research Perspective 86
4.5 Summary 89

5. An Overview of Literature 91
5.1 The Evolution of the Concept of Development 91
5.2 The Role of Communication in Development 93
5.3 Research Questions 96
5.4 Summary 101

6. Methodology 103
6.1 The Qualitative Research Paradigm 103
6.2 Data Collection and Data Analysis 106
6.3 Ethical Considerations 115
6.4 Reliability and Validity 116
6.5 Reflexivity 117
6.6 Summary 118

7. Rural China as a Testing Ground 121
7.1 An Overview of Rural Issues 122
7.2 China's Experiences of Mobilizing Communication for Rural Development 127
7.3 The Development of the Internet 137

8. Yellow Sheep River Town 143
8.1 The Profile 143
8.2 The Change Agency 145
8.3 The Internet and Education 147
8.4 E-Commerce, Manager Hu and Social Support 150
8.5 The Experiences of Yellow Sheep River and their Implications 158

9. Jinta County 165
9.1 The Profile 165
9.2 The Change Agency 167
9.3 The Diffusion of the Internet 168
9.4 The *Herald* 172
9.5 The Experiences of Jinta and their Implications 179

10. Tongnan County...185
10.1 The Profile...185
10.2 The MOST-UNDP Project...188
10.3 The Internet Use...192
10.4 The Issue of Sustainability 198
10.5 The Experiences of Tongnan and their Implications.........200

11. Pinggu District 203
11.1 The Profile...203
11.2 Taoyuan Village 204
11.3 Nandingfu Village..209
11.4 The Experiences of Pinggu and their Implications 212

12. Shouguang City and Huanglou Town.......................................215
12.1 The Profiles 215
12.2 The Diffusion of the Internet..216
12.3 The Stories of Three Rural Entrepreneurs.........................219
12.4 The Rise of Rural Entrepreneurs and its Implications.......228

13. Discussion and Conclusions..233
13.1 The Process of Internet Diffusion and Rural Development...233
13.2 Theoretical and Practical Contributions 254
13.3 Conclusions 264
13.4 Limitations of the Study and Directions for Future Research 269

Bibliography 271

Preface

The Internet came to China at a time when the country was caught in a classic paradox in the process of modernization in the developing world: an impatient desire to use whatever means possible to develop the economy fast in an effort to catch up with the West, and a strong imperative to maintain the status quo in the overall political and social structure that inherently restricts the diffusion of communication innovations. Its diffusion and development in the country have since been wrought by the contestation between those two paradoxical forces.

Politically, the Internet found itself encountering a socialist system that was inherently resistant to new communication innovations because of the fear that they would open the door for undesirable political, cultural and moral ideas – a fear derived from the traditional belief in the magic power of the communication media, the long-time logic of management in a planned economy that reacts to anything with the temptation of exercising administrative controls, and the crisis in the mandate to rule and the lack of confidence in the wake of the decline of the Communist movement worldwide.

This political concern was particularly evident in the early stage of the development of the Internet in China, which started in 1987 when professor in Tsinghua University sent an email message via the Italian Public Network ITAPAC. From 1987 to 1989, several institutions, such as Tsinghua University and the Chinese Academy of Sciences, experimented with connecting their computer networks to the Internet. In May 1989, Chinese Research Network (CRN) successfully hooked up to the Internet through the German Research Network. It was not until 1994 when the Ministry of Post and Telecommunications signed an agreement with Sprint did China seriously embark on the development of the Internet in a more systematic way. Even then, strict restrictions were exercised on the content of the Internet and access to it. In a span of 20 years, the government formulated about 11 specific regulations and laws to tame the Internet.

Economically, the Internet met a society that was in a rush to develop and reached the stage of a fascinating economic take-off. Amazed by its reach and lured by its often exaggerated potential to drive a booming e-economy, the Chinese government encouraged and invested a great deal of resources on the development of the infrastructure and hardware of the Internet. From the "year of government going on line" campaign to the construction of large-scale fiber-optic networks across the country, the government muscled its strengths to jump start the "driving engine" of an e-economy. And the results were striking. By June 2006, there were 123 million Internet users (198 times over the number in 1997), 54.5 million computer hosts, 3 million domain names, and 788,400 websites. Those figures ranked China the second largest Internet-using country in the world. On the business side, such alien Chinese names as Sohu, Sina, Baidu and Alibaba became household names and major public-holding companies in the world, attracting millions of dollars on the global stock markets.

The popularity of the Internet in urban China, the business success of the leading Internet companies in the country and the inherent imperative to push innovations from top down in a quasi planned economy naturally lead to this question: Can the Internet be used to boost the economic development and social changes in the rural areas in the Chinese context of a contestation between political and economic forces, as in the case of rural wired broadcasting networks in China in the 1960-70s, the radio math project in Nicaragua, and the media-aided "Masagana 99" campaign in the Philippines?

This is what Dr. Jinqiu Zhao set to explore in a massive study of five rural counties in China. Based on a conceptual framework that focuses on the social structure and the entire environment for the adoption of new innovations – rather than on the individuals who are held solely responsible for the adoption and rejection of innovations in the passed "dominant paradigm", Dr. Zhao attempted to examine the interplay of social/structural forces and individual factors in the diffusion of the Internet and its impact on rural economic development. She interviewed scores of farmers, extension workers and government officials, lived the experience of farmers as a participant observer, and dug into a well of often highly guarded documents. As a result, she was able to collect a large amount of vivid and first-hand qualitative data and provided a good insight into the working of the Internet in rural China. Of particular interest to the field of communication and national development were such findings that the diffusion of the Internet in rural China relied heavily on government's planning, support and financial sponsorship; that such diffusion differed from one place to another depending on the nature of the local agricultural production and business; and the diffusion of the Internet did not lead to noticeable social and economic changes – which was not a surprise given the short span of the observation taken and the initial stage of the diffusion of the Internet in those rural areas.

To students of communication and national development, reading such a book may feel like taking a journey back to the future, accompanied by a keen and lingering reminiscent sense of "been there and done it". However, readers can be assured that although the journey may feel reminiscent, the future is different – and "cyber".

Zhou He
City University of Hong Kong

1. Introduction

The massive processing power of information and communication technologies (ICTs) has fundamentally transformed the traditional ways of storing, processing, and disseminating information, which in turn creates new possibilities in generating economic growth, increasing life span, promoting universal primary education and enabling informed participation. At the same time, the great potential of ICTs poses fresh challenges – the majority of the world's inhabitants, especially those in the developing countries, have been largely excluded from the digital revolution and deprived of the opportunity to gain benefits (IWS, 2006). How ICT innovations can reach poor people and better their lives remains a challenge to every developing nation that hopes to utilize digital technologies for its development goals.

1.1 Purpose of the Study

This study aims to explore the ways in which the Internet can be used to promote social and economic development in China, focusing primarily on its rural areas, most of which are yet to benefit from advanced technological innovations. In particular, it seeks to examine how the Internet is diffused and used in the rural areas, conditioned by the political, socioeconomic and infrastructure environments; and what potentials it may have in helping rural people improve economic conditions, education, healthcare, gender equality and political participation.

1.2 Rationale and Scope of the Study

During the three decades (from the 1950s to the 1970s) when development communication research was at its prime, communication scholars failed to establish a unified theory of applicability (Okigbo, 1985). The different approaches championed by the main paradigms towards the concept of development, the role of communication in development and the solution to underdevelopment were mainly normative and assumptive in nature rather than empirically verified (Melkote & Steeves, 2001; Okigbo, 1985). Moreover, under the heavy influence of the positivist approach, the quest for generality and causal relationships between communication and development variables dominated most of the earlier studies. This led to the neglect of structural constraints that the developing countries were confronted with and an overemphasis on the role of communication and media technologies (Jayaweera, 1987; Melkote & Steeves, 2001).

Despite the absence of a well-established theory and the inconclusive empirical findings in the field of development communication (Okigbo, 1985), policy makers and researchers have never given up the hope of using communication technologies to help the developing countries bridge their gaps with the industrialized world. We have witnessed a resurgence of the modernization paradigm with the advent of new ICTs, especially communication satellites and the Internet. These technological innovations seem to have given the developing countries new hopes because they tend to be effective in bypassing geographical, economic, political and cultural boundaries. It is believed that the gaps between the developing and the developed countries could be reduced once such boundaries were bypassed. For instance, World Bank (1995) highlighted the danger of not participating in the development of ICTs in addressing African issues:

> The information revolution offers Africa a dramatic opportunity to leapfrog into the future, breaking out of decades of stagnation or decline. Africa needs to seize this opportunity, quickly. If African countries cannot take advantage of the in-

> formation revolution and surf this great wave of technological change, they may be crushed by it (¶. 3–4).

The information revolution enabled by microelectronics and digital technologies has given rise to widespread speculation and debate on the impact of ICTs on human development. ICTs allow people not only to collect, process, and transmit enormous quantities of information at reduced cost, but also to communicate with one another (Crede & Mansell, 1998). Human development centres on issues of poverty, education and health. It is argued that the rapid diffusion of ICTs facilitates human development through breaking barriers to knowledge, participation and economic activity. ICTs are considered vitally important tools for improving living conditions, education and public health through the generation of productivity gains, the expansion of educational opportunities and the promotion of medical knowledge (UNDP, 2001a; UNDP, 2004a).

With the fast development of ICTs, both the industrialized and developing countries have witnessed great social and economic transformations. Increasing numbers of governments and industries in the developing countries have invested heavily in the construction of national information infrastructures, hoping to attain their development goals thereby (Mansell & Wehn, 1998). Proponents argue that ICTs can contribute to human development in multi-dimensions (UNDP, 2001a; UNDP, 2004a). The employment and economic benefits generated by the diffusion of technologies can make direct contribution to the eradication of poverty and hunger. Moreover, ICTs can empower the poor by helping them acquire literacy and skills to escape poverty (UNDP, 2001a; UNDP, 2004a).

Despite the general optimistic views about the economic and social potentials of ICTs, it is still being debated whether the benefits of ICTs outweigh the cost of access. Whereas many evidences show that "developments in communications have produced tangible benefits for large numbers of ordinary people" (Clarke, 2004, p.iii), other experiences reveal uneven diffusion of technologies, deepened disadvantages and deprivation of equal opportunities (UNDP, 2004a). Despite the seemingly global trend of economic growth and rise in living standards, “poverty, hunger and malnutrition, ill health and disease, illiteracy and unequal access to skills and knowledge” still prevail in many parts of the developing world (UNDP, 2004a, p.1).

Some critics are sceptical about the role of ICTs in poverty alleviation (Brown, 2001; Heeks, 1999; Kenny, 2003). Heeks (1999) argued that ICTs acted mainly as communication technologies rather than as information-processing or production technologies. Others contended that investments in basic education and healthcare would address the problems of the poor more directly than providing them with access to the Internet (Chowdhury, 2000; UNDP, 2001b). Furthermore, ICTs require broadly based technical and social capabilities if they are to be used effectively. If these capabilities are not developed, it will be difficult for individuals, organizations or the whole nation to participate in the opportunities that the spread of ICT applications is making available (Crede & Mansell, 1998).

Therefore, it is important to recognize that the benefits of ICTs have not been evenly distributed despite their great potentials for development. Many sections of the developing societies, such as rural and remote communities, have been barely touched by the technological innovations. To make things even worse, there are some people, especially in the least developed countries whose lives have been negatively affected by the exclusion from the information society, with the pre-existing socioeconomic, gender and urban-rural inequalities further exacerbated as a result of the unbalanced access to information.

The widening digital divide between the industrialized and developing countries as well as between the urban and rural areas has prompted the study to examine the ways by which the Internet can be utilized to empower the rural people and improve their living standards. Owing to a lack of micro-level studies to reflect the actual contributions of the Internet to economically underdeveloped regions, there is a marked divergence between the romanticized claims about the Internet and empirical evidences that have lent support to these claims (Quibria & Tschang, 2001).

China has presented a unique testing ground for ICT-enabled development. Though China has been enjoying the fastest sustainable economic development in human history over the past decade, it has also witnessed the sharpest polarization of wealth and poverty and increased urban-rural inequalities in recent years.[1] As the existing socioeconomic and infrastructure environments in the rural areas can hardly sustain such an advanced technology as the Internet, the rapid diffusion of the technology since the late 1990s has been confined to the city centres, largely bypassing the majority of the population residing in the rural areas.[2]

Though the Internet penetration rate remains extremely low in rural China, quite a number of small scale and experimental applications of the Internet have been developed in several places under the initiatives of local governments, private companies or daring entrepreneurs. A close examination of these sporadic attempts can help to extend the knowledge of the diffusion process of the Internet as well as the possible linkage between the Internet and rural development in the contexts of rural China.

1 With the per capita annual income of rural people reaching RMB 2,936 (U.S.$ 367) in 2004, China has become one of the countries with the largest urban-rural divide in the world. The average per capita income of urban residents was 3.2 times that of rural dwellers, much higher than the corresponding figure (1.7 times) in 1984 (Lu, 2002; NBSC, 2005).

2 The total number of Internet users in rural China was about 1.5 million, accounting for 1.4% of the total Internet users and 0.12% of the population by December 31, 2005 (CNNIC, 2006).

1.3 Research Questions

Based on the gaps in research identified in the literature review (see Chapter 2, 3, 4 and 5), this study sets out to answer three main research questions:

RQ1: How are Internet diffusion and usage shaped by the political, socioeconomic, and technological contexts of rural China?

RQ2: What implications do Internet diffusion and usage have upon various aspects of rural development?

RQ3: How do governments, the private sector, as well as individuals contribute to Internet diffusion and usage? What are the possible ways of providing a viable Internet service to rural areas?

1.4 Brief Description of Methodology

Because of the exploratory nature of the study, a qualitative approach was adopted to seek an in-depth understanding of the diffusion process of the Internet and its implications for rural development. By getting herself immersed into the rural life and interacting with Chinese farmers, the researcher was able to observe and record the Internet usage patterns and appreciate the meaning of the Internet in the perception of rural users.

In accordance with the dominant players in initiating the extension of the Internet, five rural places situated at western, central and coastal regions of China were selected to generate meaningful comparisons. The varying socioeconomic conditions had also been considered as China is a country so vast in territory and population. The maximum variation sampling enabled the social phenomenon of Internet diffusion and usage to be examined under various rural contexts and helped to enrich the findings greatly.

The field studies were conducted from April to August 2005, with in-depth interviews, participant observations and documentary analysis as the main data collection methods. At each rural place, face-to-face interviews were arranged with government officials in charge of the Internet projects, the information technicians, as well as ordinary farmers. Detailed field notes were taken to record the Internet access modes, usage patterns, frequently used functions and websites, and the ways by which the Internet could possibly influence the lives of rural people. In addition, public records as well as personal documents were collected to supplement the data gathered through interviews and observations.

The qualitative data analysis involved the stages of transcribing interviews and field notes and identifying and describing categories and themes. Finally, the themes were integrated into a conceptual schema to offer a logical interpretation of the phenomenon under study. A more detailed description of the research methods is presented in Chapter 6.

1.5 Significance of the Study

This study has theoretical and practical significance for the field of development communication. First, this study departs from the positivist approach that has dominated previous development communication studies and seeks an in-depth understanding on the social and economic implications of communication technologies. Instead of treating communication and development as discrete variables, the study views the diffusion of the Internet as a social process, which is conditioned and shaped by the existing political, socioeconomic and infrastructure contexts of rural China. The qualitative data collection methods enable the researcher to observe and document the daily experiences of rural people and intuitively understand their perceptions of the Internet. The naturalistic approach also helps to identify social factors that have so far been neglected by quantitative studies and, thereby, provides alternative perspectives on and interpretations of the role of communication technology in development.

Second, the study can enrich the extant understanding on the diffusion and adoption of ICTs through the examination of the social process of Internet diffusion and the possible implications of the technology for rural development. The diversified Internet diffusion initiatives allow the study to examine the applicability of the diffusion of innovations model to the unique contexts of rural China. Whereas previous diffusion studies mainly focus on individual adoptions as a result of optional decisions, this study seeks to examine the diffusion and adoption of the Internet under the joint influences of structural and individual factors.

Unlike the extension of agricultural innovations such as new fertilizers or breeds of seeds and the diffusion of old ICTs such as radio and television, the development of the Internet involves more stringent prerequisites for access, cost and level of education, which can hardly be met by the rural areas. The underdeveloped infrastructure, widespread computer illiteracy and high costs of computer hardware and connectivity have placed the Internet out of reach of the majority of Chinese farmers. Therefore, the diffusion and adoption of the Internet are unlikely to be determined very much by the will of the individuals.

Moreover, the Chinese cases may add to the conventional understanding of the concepts of innovation adoption and usage. Given technical and knowledge gaps between the Internet and rural people, efforts have been made to circumvent the barriers posed by the technology and facilitate Internet access. The available experiences of Internet applications may generate new interpretations of Internet adoption and usage, which vary greatly from those taking place in urban settings or as a natural step in the rural economic and infrastructure development.

Third, as there has been a lack of micro-level research and empirical evidence to verify the perceived efficacy of the Internet in reducing poverty and improving the living standards of rural people, the study is a step closer towards a clear understanding of the ways by which the Internet can be accessed and used by rural people and the actual contributions of the technology to various aspects of rural development.

On the practical side, the findings of the study have important implications for narrowing the digital divide between the urban and rural areas in China. The innovative practices of Internet diffusion and usage across the rural places may offer practical solutions to the provision of Internet services or the transmission of online information in the rural areas where technical and knowledge gaps need to be bridged. The identification of the possible ways of keeping sustainable rural Internet services not only helps to make policy recommendations for governments and development planners at national and local levels but also provides valuable lessons for other developing countries with similar development concerns.

1.6 Outline of the Book

There are thirteen chapters in this book. Chapter 1 discusses the purpose and rationale of the study, the research questions, the research methods and analytical approaches used, and the significance of the study.

The literature review runs from Chapter 2 to Chapter 5. Chapter 2 traces the origins of three major paradigms in the field of development communication, reviews and critiques their conception of development and the role of communication in the development process, and assesses their contributions.

Chapter 3 reviews and discusses the concept of human development, the architecture and technological characteristics of the Internet, and the ambivalent attitudes among scholars regarding the possible contributions of ICTs towards various aspects of rural development.

Chapter 4 reviews the main concepts of the diffusion of innovations model and the available findings. It also analyzes the limitations of the previous diffusion studies and identifies the understudied areas.

Chapter 5 synthesizes the theories, views and research findings covered from Chapter 2 to Chapter 4. It examines the evolution of the concept of development and the extant findings on the role of communication in development. The knowledge gaps identified in the literature help to formulate more focused research questions to address these gaps.

Chapter 6 provides a detailed description of the methodology for this study, including the specific methods used in the collection of data, the selection of the research settings, the sampling procedures, and the techniques used in data analysis.

Chapter 7 gives an overview of China's experiences of mobilizing communication for rural development as well as some background information about the diffusion of the Internet in the country. It provides a contextual understanding of the political, socioeconomic and infrastructure environments of rural China.

Chapter 8 to Chapter 12 present the major findings of the study by providing detailed descriptions of the diffusion and usage of the Internet at the five rural places. Chapter 13 discusses the process of Internet diffusion and the implications of the Internet for rural development. The conclusions underline the theoretical and practical contributions of this study, address the limitations, and identify directions for future research.

2. Development Communication

Central to the studies of development communication are the nature of development and the role of communication in the development process. The rising of the modernization paradigm, the dependency school as well as the pluralist perspective has made great contributions to the evolution of the development concept as well as our understanding of the roles accorded to communication in development.

2.1 The Modernization Paradigm

Throughout the 1950s and the 1960s, there was widespread optimism about development communication among scholars and development planners. The intellectual discussions about development as well as the national development programmes of many developing countries in Asia, Africa and Latin America were shaped by the modernization paradigm, often referred to as the dominant paradigm.

2.1.1 The Conception of Development

a) Modernization as Social Evolution

During the post World War II period, more than 40 newly independent nations emerged in Asia, Africa and Latin America, winning independence after centuries of colonial rule (AED, 1985). Being ill-equipped for independence, these newly independent countries continued to be plagued by poverty, unemployment, famine, overpopulation and economic underdevelopment. How to make these newly-emancipated societies catch up with the industrialized nations and attain the same kind of economic growth became a major concern of government officials, development planners and academicians (Melkote & Steeves, 2001).

Social evolution theories, which originated from Darwin's evolutionary concept, had been applied to the explanation of the development process of human societies by social theorists like Herbert Spencer and William Sumner. They believed that all societies, like biological organisms, developed from simple, small-scale beginnings into more complex industrial and post-industrial societies. This development process was thought to be unilinear, that is, there was one line of development from simple to complex (Melkote & Steeves, 2001; Rostow, 1971).

The social evolution theories, based on a mere description of the development pathway of Western Europe and North America, prescribed that the developing countries should undergo the same stages of social transformation to achieve social and economic development. The unilinear development perspective was clearly illustrated in Rostow's *The Stages of Economic Growth: A Non-Communist Manifesto* (1971). Rostow held that societies should be classified into five categories according to their different degrees of development in economic dimensions: the traditional society, the preconditions for take-off, the take-off, the drive to maturity and the age of high mass-consumption.

The same dynamic of social development from a traditional to a modern society was advocated by Daniel Lerner in *the Passing of Traditional Society: Modernizing the Middle East* (1958). Lerner also emphasized that the institutional and psychic changes which had taken place in the West were essential to the modernization process. According to Lerner, the Western model of modernization had set an example for the rest of the world to emulate. He described the process of social change thus, "Everywhere, for example, increasing urbanization has tended to raise literacy; rising literacy has tended to increase media exposure; increasing media exposure has 'gone with' wider economic participation (per capita income) and political participation (voting)" (p.46).

As a direct extension of the social evolutionary theories of the 19th century, the modernization paradigm conceived the Western development model as a universal path that could not be bypassed and excluded other alternatives to development.

b) Modernization as Economic Growth

The five stages of economic development proposed by Rostow (1971) were based on the development route of Western European countries. Economists and development planners in the developing countries took economic growth achieved through industrialization as an essential pathway to development. The effective cure for underdevelopment was to replicate Western models of development such as urbanization and the introduction of advanced technologies (Rogers, 1976). Consequently, the problem of development or underdevelopment was basically defined by and limited to economic terms.

GNP or per capita income became a main indicator and measurement of national development. The developing countries were advised to invest heavily in the industrial sector at the expense of agriculture and other social development. The social and cultural dimensions of development were not stressed because such dimensions could not be quantified by dollar values (Singhal & Rogers, 1989).

The quantification of development could also be attributed to the positivist approach employed by most Western scholars in the 1950s and 1960s (Rogers, 1976; Singhal & Rogers, 1989). This research outgrowth led to equation of development with the social gains that could be measured and quantified in numbers. The modernization advocators believed that the economic development process of Western nations could be generalized to the developing countries if the preconditions for development were met (Obijiofor, 1994; Melkote & Steeves, 2001).

c) *Modernization as Individual Change*

While some scholars conceptualized modernization from a macro-level of social and institutional evolution, theorists such as Hagen, McClelland, Lerner and Rogers sought to define and analyze modernization from a micro perspective. They expected to establish psychological and social prerequisites of development through changing individuals' internal determinants of behaviours such as their values, attitudes or personalities (Obijiofor, 1994). Hagen's (1962) concept of "withdrawal of status respect" argued that the transformation from a traditional economy into a prosperous modern one should rely on innovative personalities that the population of the developing countries aspired to. McClelland's (1961) notion of "achievement motivation" held that need for achievement among individuals was the major psychological factor that accounted for the advance of some nations and the backwardness of some others.

Weber (1958) considered cultural norms and values as the most important elements accounting for the emergence of capitalism and its absence. The development of capitalism in Western Europe was attributed to the Protestant ethic of asceticism and its nurture of the spirit of capitalism. According to Weber, the spirit of modern capitalism involved the incentive to accumulate capital, make profit and invest in productive endeavours that would lead to further acquisition of wealth.

Lerner (1958) attached great importance to individuals' psychological dimension in social transition because he believed how individuals solved their personal problems and lived their private lives constituted the whole changing process. Modernization was the transition from non-participant to participant, with empathy as its distinctive personality mechanism. According to Lerner, a modern society was made up of high empathic individuals who were participant in societal affairs while a traditional society was characterized by constrictive personality. Empathy, "the capacity to see oneself in the other fellow's situation", became an indispensable skill for people to move out of traditional settings and operated efficiently in a changing world (p.50).

Rogers and Svenning were also among those who stressed the importance of individual change in values and attitudes in order for modernization to occur. Based on their study of peasants in Colombian villages in the Andes Mountains, Rogers and Svenning (1969) commented that it was important to modify the attitudes and behaviours of peasants and help them foster pro-development attitudes and modern psychic traits.

2.1.2 The Role of Communication

The central assumption of the modernization paradigm was that communication, by spreading knowledge, disseminating information and stimulating desires for change, could become a very powerful and direct force for development (Goonasekera, 1987; Jayaweera, 1987). Accordingly, scholars and theorists held a very optimistic view of what communication could contribute to the modernization process. Mass media communication was thought to have the power to break down the traditional societies and help to establish a modernized nation (Rogers, 1976).

The 1950s and 1960s witnessed an expansion of mass media among most developing countries. The improved literacy rate led to greater exposure to the print media and transistor radios penetrated into even remote villages. The popular belief was that mass media could convey pro-development content from government development agencies to the mass audience. Communication process was built upon a linear and cause-effect model (Singhal & Sthapitanonda, 1996). Lerner (1958), Rogers (1983), Schramm (1964) and Inkeles and Smith (1974) were the influential advocators of the modernization paradigm at the time.

Lerner (1958) considered individuals as key elements to social change. People must change in order for a society to get transformed. Therefore, modern society should be composed of modern men and women. The Traditionals need acquire empathy to become participant as required by the modernization process. Exposure to the mass media was a crucial means for the Traditionals to develop the modern psychic trait. Lerner expounded the point thus,

> Radio, film and television climax the evolution set into motion by Gutenberg. The mass media opened to the large masses of mankind the infinite *vicarious* universe. Many more millions of persons in the world were to be affected directly, and perhaps more profoundly, by the communication media than by the transportation agencies (p. 53).

Lerner (1958) argued that people's experience through physical or graphical mobility was expanded and multiplied by "mediated experience through mass communication" (p.52). The media enlarged a person's horizon of the world by increasing his empathic capacity to imagine himself in different situations. He implied that as a substitute for personal experiences, mass media could enrich people's understanding of the outside world and help them develop the essential aptitude of a modern man. Lerner commented that only an urban, literate and democratic society could sustain continuing social changes into existing institutions and "media participation" had a lot to contribute to the creation of an environment favourable for modern and democratic nations (p.57).

Moreover, Lerner (1958) considered a communication system both index and agent of social change. He contended that as mass media only prospered in modern societies, they could serve as an important indicator of modernization. Lerner used urbanization, literacy, voting and media participation as indices of modernity (public participation). He hypothesized that participation in the four sectors could represent the whole process of modernization in a social system.

Rogers and Svenning (1969) emphasized the essential role of communication in the modernization process through tracing how new ideas were adopted in rural communities to bring about changes. They defined modernization as "the process by which individuals change from a traditional way of life to a more complex, technologically advanced, and rapidly changing style of life" (p.48). Rogers and Svenning argued that the transition from a traditional way of life to a modern life style was based on communication and adoption of new ideas.

Without introducing new ideas from external sources, it was hard for an ill-informed village to come up with any invention. Therefore, mass media as well as interpersonal communication became vitally important in disseminating development messages and securing positive changes.

In *Diffusion of Innovations*, Rogers (1983) pointed out the importance of two-step flow of communication in development. The information and knowledge disseminated from the mass media was usually first adopted by opinion leaders and then passed on to the public. Mass media had an irreplaceable role in its capability of reaching the mass audience, whereas interpersonal communication exerted influence on the individual's decision on adopting or rejecting new ideas or practices. Rogers argued that communication would achieve a multiplying effect in serving the end of modernization if mass media and interpersonal communication were combined together.

Schramm (1964) took sides with Lerner and Rogers by proposing that social change occurred both at the system level as well as the individual level. Aggregation of individual improvements could lead to structural transformations. Schramm argued that mass media were agents of social change involving "the transition to new customs and practices and, in some cases, to different social relationships" (Schramm, 1964, p.114).

Schramm (1964) held that communication was entrusted with three tasks to fulfil in national development – the informing function, the decision-making function and the teaching function. First, mass media should supply the public with messages about national development. Through exposure to mass media, the public could get a better understanding of the needs as well as the means of social changes and hence their aspirations towards development were enhanced. In this respect, mass media were most effective in fostering a pro-development environment. But it was not the case in the decision-making process. Mass media could only play an indirect and facilitating role to alter the deep-held values and practices. Instead, channels of interpersonal communication were far more influential in bringing about attitudinal change as group leaders played a crucial part in setting up development norms among the public. As to the third task of educating the public, Schramm attached great importance to the mass media because of its efficacy in spreading knowledge and information.

As aforementioned, the central thesis of the modernization paradigm revolved around the role of communication in effecting social change. The high hope pinned upon development communication turned out to be apparent failures in many developing countries that had been engaged in practicing the dominant model of development at that time.

2.1.3 Critique

The dominant paradigm together with its underlying assumptions was widely accepted by many developing countries without questioning its validity until the 1970s. The modernization thesis was under severe attack as the much prophesied "take-off" did not take place. The underdeveloped countries did not experience the kind of development as predicted by the Western scholars. Instead, serious social problems emerged under the misguided conception of development as economic growth.

The ethnocentric nature of the dominant paradigm, its "*internal blame* bias", the overall failure of "big projects" guided by the top-down model and the negative social consequences incurred, and the exaggerated role of communication in the development process became subject to strong criticism (Melkote & Steeves, 2001, p.90).

First, the dominant paradigm had much to blame for its Western ethnocentrism. Mowlana and Wilson (1990) pointed out that the hidden agenda behind the modernization approach aimed at spreading the Western ideologies and value systems to developing countries. Sachs (1992) also commented on the ethnocentric nature of the dominant paradigm, "there are lead runners who show the way to latecomers; it suggests the advancement is the result of planned action" (p.6). To many Third World countries, modernization was equated with Westernization.

Though the Western model of development, which placed a high value on science and technology, modern forms of social institutions and democracy, proved to function quite well in the developed countries, it did not produce the same appealing results in the developing countries. The mechanical application of the economic and political systems of the West could not guarantee the rise of living standards and other positive social changes among the developing countries. Although the adoption of capital intensive technologies helped to improve efficiency and output, it also resulted in serious social problems in the developing countries where labour was abundant. The concomitant unemployment and displacement of workers caused a deluge of urban slums owing to large-scale migration of rural surplus labour to the cities (Jayaweera, 1987; Singhal & Rogers, 1989). The lack of compatible structural reforms to match the introduction of capital intensive technologies led to the disappointing experiences in the developing countries.

Another side effect brought by the transference of capital-intensive technologies was the increasing dependence of the national governments of the developing countries on international loans and multinational corporations under the control of the developed nations. The developing nations soon realized that political independence could not guarantee economic independence, and their dependence on the industrialized nations became stronger than ever (Rogers, 1976).

Second, the misguided conception of development as a higher standard of living quantified by indicators such as per capita income, per capita consumption of resources and GNP intensified the existing social instability. Developing countries with impressive rates of growth did not achieve much progress in reducing unemployment, social inequality and poverty during the so-called Development Decade in the 1960s and in some cases the situation had deteriorated. Weaver and Jameson's (1978) study, based on available data on several developing countries such as Kenya, Brazil, Ecuador and Turkey, suggested that the economic planning guided by the modernization paradigm produced mixed results. On the one hand, high rates of growth were accompanied by increasing unemployment; on the other hand, a very small fraction of the population controlled a large share of national income. Other studies showed that despite the economic expansion, the income of bottom 40% of the people in certain developing countries fell not only in relative but also in absolute terms (Melkote & Steeves, 2001).

Third, the dominant paradigm held that the causes of underdevelopment were internal rather than external to developing countries. The modernization theorists often sought to attribute the backwardness of developing countries to their traditional social conventions, values and their impoverished, illiterate population. The external constraints such as centuries of colonial rule and the economic dependence of these countries on multinational corporations and international aid programmes were either downplayed or totally ignored. As a result, the dismantlement of traditionalism, which served as a prerequisite to the establishment of modernization, led to the total neglect of the historical and cultural context that development should be situated in.

Fourth, the causative role of communication assigned by theorists in terms of its relationship with development proved to be short of empirical evidence. Though a series of studies were conducted in some developing countries (such as Colombia, Turkey, India and so on) with an attempt to prove that mass media were important causative factors conducing to modernization, the findings only demonstrated various degrees of correlation between the two variables rather than the former caused the latter. Hence, the powerful role that mass media played in development was assumed rather than empirically established (Rogers, 1976).

The limited and biased notion held under the modernization paradigm about the nature of development and the role of communication in development did not fit the reality and potential of the developing countries. Failure of many development programmes prompted scholars and government officials to seek alternative explanations for the causes of underdevelopment.

2.2 The Dependency Paradigm

The failure of the programme of the UN Economic Commission for Latin America (ECLA) in the early 1960s triggered the rise of the dependency paradigm. Whereas the modernization paradigm attempted to examine development from the viewpoint of the United States and other Western countries, the dependency approach viewed the issue of development from the perspective of the developing or peripheral countries.

According to the dependency paradigm, development and underdevelopment should be analyzed within the context of international system of economic relations, which in the present phase of capitalism split the world into two groups – the core and the periphery. The core countries were industrialized and rich whereas the periphery were underdeveloped and exploited (Goonasekera, 1987). The dependency school contended that the periphery countries' underdevelopment and the core countries' development were "two aspects of a single process of capital accumulation, leading to regional polarization in the global economy" (So, 1990, p.104).

2.2.1 The Conception of Development

As the dependency perspective was raised as an antithesis to the modernization paradigm, its conception of development and the role of communication often took a stand against the modernization school to challenge the validity of Western model of development. Frank (1967, 1969) and Dos Santos (1971) played leading roles in expounding the fundamental concepts of the dependency school.

Frank (1967, 1969) authored the concept of "development of underdevelopment" and the theoretical model of metropolis-satellite. Frank argued that it was unfair to attribute the backwardness of the developing countries to traditionalism and primitiveness as many developing countries were under the Western colonial rule for over a century and colonialism had actually altered their development course. Frank used the concept of "development of underdevelopment" to explain that the development of Western countries was actually the direct cause of the widespread underdevelopment in the Third World.

Frank (1967, 1969) formulated the metropolis-satellite model to illustrate the relationship between the developed and the underdeveloped. The model derived from the relationship between the cities in the developing countries and the Western metropolis during the colonial period. The colony governors set up new cities in the developing countries to facilitate the transfer of "economic surplus" (in the forms of natural resources, commodities and profits) back to their home countries (So, 1990, p. 97). The prosperity of Western metropolis was built upon plundering the satellite cities in the developing countries. Frank assumed that the prototype of the relationship formed during the colonial period could be generalized to the contemporary exploiting nature between the developed and the underdeveloped worlds.

Whereas Frank (1967, 1969) attempted to offer explanations to the possible causes of underdevelopment, Dos Santos, another leading figure in the dependency school, focused on analyzing the dependent relationship between the industrialized and the underdeveloped.

Dependence, according to Dos Santos (1971), meant the relationship between two or more countries "assumes the form of dependence when some countries (the dominant ones) can expand and can be self-starting, while other countries (the dependent ones) can do this only as a reflection of that expansion" (p.226). Dos Santos pointed out that the relationship between the dominant and dependent countries was unequal in nature because development of the former was based on the exploitation of the latter.

2.2.2 The Role of Communication

Under the dependency paradigm, the main thesis regarding communication centred on media or culture imperialism. The modernization paradigm took the nation as the frame of reference as far as communication was concerned, whereas the dependency paradigm focused its analysis on the information flow at the international level. The underlying assumption of media imperialism was that the messages conveyed through the Western media would expose the people of developing countries to the capitalist ideology and consumer culture and inculcate Western beliefs and values into these people (Servaes, 1991).

The dependency paradigm claimed that the free flow of information advocated by modernization theorists only favoured those countries commanding greater communication resources. These privileged countries were accused of using liberalism as a tool for economic and ideological expansion. The Non-Aligned Movement opposed the Western liberal concept by arguing that communication should be two-way rather than one way (from the centre to the periphery) and it should be free as well as balanced.

With the UNESCO debate over new global economic and information orders, development was seen as the establishment of true sovereignty and independence from the West. In terms of communication, it referred to the severing up of mass media in the developing countries with the existing dominant system and rejecting the Western cherished beliefs and values (Stevenson, 1988). The debate over the New World Information and Communication Order (NWICO) emphasized more on the political role of mass media both in the developing countries and the Western nations, in contrast with its role of promoting social and economic development under the modernization model.

2.2.3 Critique

Though the dependency paradigm emerged as an antithesis of the modernization paradigm, there was little primary difference between the two as far as the content of development was concerned. Both perspectives took economic growth as their major development concern and applied the quantitative methods to the study of development and underdevelopment (Servaes, 1991).

The dependency school was criticized for its concentration on the causes of development and underdevelopment rather than on building the concept of development itself (Mowlana & Wilson, 1990). Therefore, the dependency school was often not taken as a new model of development but a combination of critiques targeted at the modernization paradigm.

Servaes (1991) argued that the studies on the imbalanced flow of information at the international level had been confined to only one medium or dimension, such as the most-talked-about international news. One of the weaknesses of the dependency approach lay in that many of the dependency oriented debates and studies were under the bias of nationalism and ideology and the same thing could be said about its communication aspect. These predispositions in turn shaped the interpretation of research data on international communication.

Regardless of the variances among nation states and their historical origins, the dependency school sought a general pattern of dependency between the core and the periphery. Servaes (1991) argued that the dependency advocates should take into account the fact that there was much differentiation among the peripheral countries due to historical reasons and these countries themselves were undergoing constant transformations. Looking at the peripheral countries as a homogeneous group was to simplify the complex situation.

Another weakness of the dependency paradigm lay in its attempt to examine the causes of underdevelopment from the outside. It was believed that the hindrance to national development was colonialism and the unequal division of labour imposed by countries at the core. The internal obstacles such as the lack of capital, technological skills and democratic institutions were not taken into account. As a result, the dependency school overemphasized the clashes at the international level and was not aware of the confrontations within the peripheral states, such as the clashes of interests between the state and media owners, government and population.

The dependency school took a radical view towards the solutions of underdevelopment. They held that it was impossible for periphery countries to secure any genuine development as long as there was constant flow of economic surplus to the core countries. They suggested that periphery countries should break up ties with core countries and adopt a self-reliance model through "relying on their own resources and planning their own paths of development so as to achieve independence and autonomous national development" (So, 1990, p.105).

2.3 The Alternative Pathway to Development

The developing countries' experiences during the Development Decade suggested that no country could be totally autonomous and self-reliant and a country's performance was often subject to both internal and external factors. The pluralist approach emerged to address the need to view development from a global perspective as more and more countries realized that they were interwoven into a world system. As a rejection and critique of both the modernization and the dependency schools, advocates of the pluralist perspective believed that there was no universal model of development and development should be seen as "an integral, multidimensional, and dialectic process which can differ from one society to another" (Servaes, 1991, p. 63). In contrast to the economic and political orientation of the previous paradigms, the new approach took a pluralist view that each country should seek a unique development pathway based on its own historical and social conditions and development needs.

2.3.1 *The Conception of Development*

The pluralist perspective took a normative rather than a positive approach as to how development and communication should be instead of how they actually are. Many scholars (Chapel, 1980; Galtung, 1980; Perroux, 1983) contributed to the formulation of the "Another Development" concept. Rogers (1976) and Servaes (1991) summarized the central components of the conception as social justice, balanced development, self-reliance, popular participation, and integration of the traditional and modern systems.

The pluralist perspective implied that the issue of development should be understood relatively, which meant that each country might have both developed and undeveloped sectors and no country could claim it was developed in all aspects. The new perspective tended to hold a balanced view as compared to the bipolar approaches preferred by the old paradigms, which saw the world being divided into traditional vs. modern or core vs. periphery. Servaes (1991) commented that, "because the previous paradigms were unable to combine economic growth with social justice, this paradigm tried to rethink the problem of freedom and justice in the relationship between man and society, and the problem of the limits to growth as inherent in the relationship between society and nature" (p.64).

The meaning of development started to emphasize the entire realm of social development. Development meant more than economic growth and the application of technological innovation – it also incorporated social equality and the reconciliation between nature and society in order to make any development to be sustainable. The emphasis of development was placed on improving the life quality of the rural and urban poor. Equal distribution of information, resources as well as socioeconomic benefits became a development priority to many nations (Rogers, 1976). Wang and Dissanayake (1984) defined development as:

> A process of social change which has as its goal the improvement in the quality of life of all or the majority of people without doing violence to the natural and cultural environment in which they exist, and which seeks to involve the majori-

> ty of the people as closely as possible in this enterprise, making them the masters of their own destiny (p.5).

The self-reliance and popular participation advocated by the pluralist perspective relied on the local initiative and resources to mobilize support from all social sectors. The new perspective of development gave full recognition to the potential at the individual and local dimensions and humanized the whole development process. It proposed that people should develop themselves rather than be treated as subjects to be developed (Rogers, 1976).

In contrast to the modernization view that the traditional values, norms and institutions should be abandoned if modernization was to take place, the pluralist perspective fully recognized the contribution that could possibly be made by the traditional practices. The pluralist advocates promoted integration between the old and new ideas. The value of the cultural and social heritages was taken into consideration when development strategies were designed and executed (Rogers, 1976).

2.3.2 The Role of Communication

The self-development model accorded communication a completely different role from the top-down development approach. It rejected the view of communication as a one-way and centralized process from the sender to the receiver. Instead, communication was seen as basically two-way, interactive and participatory at all levels (Servaes, 1991; Singhal & Sthapitanonda, 1996). The pluralist perspective "favours multiplicity, smallness of scale, locality, deinstitutionalization, interchange of sender-receiver roles, horizontality of communication links at all levels of society and interaction" (McQuail, 1983, p.97) and "rejects the necessity of uniform, centralized, high-cost, highly professionalized, state-controlled media" (Servaes, 1991, p.68).

Accordingly, mass communication played a supportive and less directive role in assisting the self-development activities of the local communities. The mass media disseminated development messages and information about innovations attuned to the local request. Moreover, the mass media could also be used to set examples for others to emulate so that neighbouring groups could benefit from the successful accomplishment (Melkote & Steeves, 2001).

Due to the proliferation of traditional cultures since the 1970s, many developing countries have witnessed a growth of dualistic communication structure combining the mass media at the national level with the traditional media at the grassroots level. The traditional media were characterized by high participation, high credibility and a strong integration with other social institutions whereas the mass media, as a transplant from the Western countries, could hardly enjoy the same penetration and popularity among the public (Servaes, 1991).

The pluralist approach argued that the Western media content as well as technologies could be utilized to serve the development purpose of the developing countries. Relying on the media technologies imported from the Western countries, communication institutions in the developing countries could be devoted to fostering an indigenous culture free of Western values and behaviour patterns. Moreover, even the imported cultural programmes could be adapted to cater to the local cultural preference. Instead of totally rejecting or severing with the Western media, the pluralistic perspective promoted a rational and dialectic approach towards the Western media and cultural hegemony (Servaes, 1991).

The pluralist perspective attached greater importance to communication at the regional and local level than that at the country level. The rationale was that the basic development needs and social and economic problems were usually discussed at the local communities. Therefore the local people should be mobilized to participate in decision-making in terms of planning and producing media content. The development planners should attend to the viewpoints of the local public and help them realize self-management (Servaes, 1991).

2.3.3 *Critique*

The 1970s witnessed the passing of the modernization paradigm as a dominant model of development in Asia, Africa and Latin America, even though the paradigm is still followed enthusiastically in some nations. Countries like Cambodia, Vietnam and Tanzania started to seek their own development route based on China's experiences of self-reliance and public mobilization.

The premise underlying the pluralist perspective was that the developing countries could hardly achieve modernization by following the model of the industrialized countries because the historical conditions under which the capitalist countries prospered were no longer there and there were no places for latecomers in the capitalist arena. Therefore, developmentalists proposed that the developing countries should adopt policies stressing basic needs, self-reliance and participation (Jayaweera, 1987).

The pluralist approach has been criticized for its idealism. Its promotion of self-reliance and decentralization of development based on the local initiatives and grassroots participation did not seem to bear much relevance to the policymakers and development planners. Jayaweera (1987) commented that only a few countries had opted for this development package. Moreover, the advocacy of participation was questioned for its Western origin. The liberal approach adopted by the pluralist perspective and its belief in individualism made many scholars doubt its applicability to non-Western cultures. Whether the public in the developing countries perceived and interpreted popular participation the same way as their Western counterparts was highly questionable (Singhal & Sthapitanonda, 1996).

The pluralist paradigm has not been widely practiced due to its utopianism. One of the difficulties in applying the pluralist approach lies in that despite the shared objectives and visions among people and institutions at different levels, they are lacking in cooperation and concerted efforts. Servaes (1991) suggested that in order for the development strategies guided by the pluralist approach to be successful, the local, regional, national and international initiatives must be combined and integrated.

2.4 Summary

This chapter traced the origins of three major paradigms in the field of development communication, reviewed their conceptualization of development and the role of communication in the development process, and assessed their relative contribution to the developing countries from the 1950s to the 1980s. The main tenets of the modernization, dependency and pluralist paradigms are summarized as follows.

Under the modernization paradigm, development was considered a uni-directional, progressive and irreversible process involving the transformation from an underdeveloped traditional society to an industrialized, materialistic and democratic modern society. The developing countries were urged to follow the development path of Western countries through introducing capital-intensive technologies and emulating the Western socio-political system. The hindrance of development was seen as internal rather than external to the traditional society. The state of backwardness was attributed to the traditional way of thinking and acting, and conservative social norms and values. The main indicator for development was GNP or per capita income.

The advocates of the modernization paradigm attached great importance to the potential role of communication in helping the developing countries achieve their development goals. Mass media communication was taken as "magic multipliers", an essential tool for spreading informative and persuasive messages from the development agencies to the public and fostering a pro-development climate within a nation state; whereas channels of interpersonal communication were seen as having a significant influence on individuals' decision-making process in terms of adopting or rejecting innovations or ideas. Consequently, the communication process was supposed to be a top-down, cause-effect model and a one-way flow of information from senders to receivers. During the process of communication and development, centralized planning was preferred in implementing development strategies.

As an alternative based on critiquing the apparent deficiencies of the modernization paradigm, the dependency paradigm adopted a politically radical view of underdevelopment. As an attack against the internal or individual bias held by the modernization scholars, the dependency paradigm analyzed development within the context of international relations. The dependency advocates argued that the extraction of economic surplus from colonized regions to the developed nations and the unequal international division of labour contributed to the widespread impoverishment in the developing countries. Therefore, the underdevelopment of the developing countries or the periphery countries was directly caused by the development and prosperity of the industrialized core countries. The prescription for the developing countries to terminate their dependency and underdevelopment was to disassociate with the core and strive for self-reliance.

Originating from Latin America, the dependency paradigm was under the heavy influence of ideology and nationalism, which shaped its perception of information flow at the international level. Cultural and media imperialism thesis severely condemned Western countries for their hegemony in the cultural sector through monopolizing communication resources and media content. With an aim to protect their nations from cultural erosion, many developing countries demanded the NWICO and NWEO. The dependency paradigm was taken more as a political outcry than a sound theory. This led to the questioning of the validity of related studies as the interpretation of research data could be contaminated by researchers' ideological bias.

The pluralist perspective emerged as an alternative to the modernization and the dependency paradigms in the 1970s. Its advocates argued that as all nations were interdependent in a global system, both internal and external factors might influence the development process. The new perspective adopted a normative approach towards the development issues – development should be need-oriented and endogenous and should promote self-reliance, dignity, equal access and popular participation at all levels. As countries differ from each other in social, economic structures and development concerns, there should be no universal model for development. Each country should design its development strategy based on its own needs and strengths.

From the pluralist perspective, the conception of development was expanded to social sectors. Development was seen as more than the accumulation of capital, and its non-economic dimensions like the social justice and the quality of life were much emphasized. The flourishing of the traditional culture at the grassroots level led to the integration of both modern and conventional media within a country. To a certain extent, the traditional media outperformed the modern media thanks to its feature of higher participation and credibility. The dualistic communication system was said to have well served the purpose of popular participation in decentralized self-development planning and execution. Communication process therefore should be two-way, interactive and participatory.

In a nutshell, the shifts in thinking in the field of development communication have provided a panorama of the evolution of the concept of development and the different perspectives on the role of communication in development from the 1950s to the 1980s. What follows next is a detailed review of the rising of the human development concept in the 1990s and onwards as well as the possible contributions of ICTs towards human development in general and the impact of the Internet on enhancing rural development in particular.

3. ICTS and Rural Development

The beginning of the latest ICT revolution was marked by the invention of "a computer on a chip" in 1969 and the reduced cost of semiconductor technologies and microelectronics (Mansell & Wehn, 1998, p.12). Over a relatively short span of around 30 years, new ICTs have been deeply woven into our social and economic fabric and become an integral part of the daily life. It is hard to imagine that the modern society can function without their presence.

The enormous potentialities and benefits associated with new ICTs such as the satellite, wireless technology, and the Internet have generated a resurgence of enthusiasm among developing countries with hopes of using technologies as "a means to transcend structural and historical weaknesses of developing nations in the economic, political and social spheres" (UNDP, 2004a, p.17). They believe that ICTs can help the developing world leapfrog the intermediate stages of technological development and participate in the global knowledge economy together with the industrialized nations.

3.1 Technological Convergence

Before the advent of microelectronics and digital technologies, ICTs consisted of three separate fields – computing, telecommunications and broadcasting (Goonasekera, 1997). Technological advances such as digitization have contributed to the convergence of three distinctive industries into the ICT sector. Therefore, convergence and informatization are viewed as two distinguished features of ICTs (Mansell & Wehn, 1998).

ICTs refer to "information-handling tools – a varied set of goods, applications and services that are used to produce, store, process, distribute and exchange information", including "the 'old' ICTs of radio, television and telephone, and the 'new' ICTs of computers, satellite and wireless technology and the Internet" (UNDP, 2001b, p.2). These tools have formed a "networked world" – "a massive infrastructure of interconnected telephone services, standardized computing hardware, the Internet, radio and television, which reaches into every corner of the globe" (UNDP, 2001b, p.2).

Old ICTs are seen as "skill independent" because old media technologies do not make high demands on users' skills, whereas new communication technologies need users to acquire some basic training and education for their applications and therefore are more "skill dependent" (UNDP, 2004a, p.2). New ICTs such as the Internet require a command of specific skills (E-literacy) and this poses difficulties to underprivileged groups or regions in terms of ICT diffusion and adoption, who have problems even with the general improvement of literacy (UNDP, 2004b).

Much of the hype associated with new ICTs for development can be attributed to the relative advantages over their predecessors. The architecture and technological characteristics of the Internet-based technologies have influenced radically the way societies communicate, organize the production and distribute the wealth. It took the Internet only four years to reach 50 million households, whereas the corresponding periods for radio and television were 25 and 13 years (Mann, Eckert, & Knight, 2000).

The Internet is defined as a "network of computer networks" – "a global network of interconnected computers that communicate freely and share and exchange information" (Massey & Baran, 2001, p.162). The Internet was originally designed for U.S. military to maintain command over its missiles and planes in case that a nuclear attack would disrupt or destroy conventional means of communication. After the development of personal and microcomputers, millions of individuals and especially the business community overtake government and academic institutions to become the chief users of the Internet.

Decentralization is innate to the Internet, and freedom and openness are central to the Internet culture. The non-hierarchical architecture of the Internet helps to break down the information hierarchy and monopoly imbedded in traditional media. Unlike radio or TV, the Internet does not operate as flows from a point or centre to a mass audience with limited feedback channels and permits the "bypassing of the strictures of formal political organizations and the scrutiny of government authorities" (Rodan, 1998, p.63). Under most circumstances users do not need permission to get online and offline and can express their opinions freely and openly (Massey & Baran, 2001).

The Internet is also a medium that integrates both mass media and interpersonal communication. Unlike the traditional media, the Internet does not operate from a centre point to a mass audience with limited feedback channels. The Internet allows multiple modes of communication: one to one (e-mails), one to many (newsgroups), and many to many (chat lists and discussion forums), which greatly enhance the reach of the media and facilitate information sending and sharing. Once a user is connected to the Internet, it is technically possible for him to connect to any other computer elsewhere on the network, wherever it is physically located (Long, 1995; Massey & Baran, 2001).

The multi-functions and operations enabled by the Internet (such as emails, remote connections, file transfer, World Wide Web browsing and E-commerce) entail great implications for the empowerment of individuals. The Internet facilitates multidirectional flows of information and permits individuals as well as minority groups to participate in political and socioeconomic activities. The development, diffusion and operation of radio and television are frequently shaped by legislators, regulators and related industries because of the limited spectrum and frequencies. Audiences do not play an active role in the one-way communication mode because they are passive receivers of programmed messages. The Internet does not have audiences but users. The development and operation of the Internet can be more determined by the will of individuals in terms of what they choose to do (Massey & Baran, 2001). The former NBC and PBS president Lawrence K. Grossman (1999) argued, "Gutenberg made us all readers. Radio and television made us all first-hand observers. Xerox made us all publishers. The Internet makes us all journalists, broadcasters, columnists, commentators, and critics" (¶.3). The technology provides much freedom for individuals to seek as well as create and send information.

Another characteristic that facilitates the easy access and wide adoption of the Internet is its relatively low cost. Communication via the Internet is much cheaper than over the telephone or fax considering the amount of information transmitted. Whereas a transatlantic call engages a telephone line between two people, emails or document files measured by megabits can travel along the same line simultaneously and instantaneously (Richardson, 1998). The innovations of microprocessors, semiconductors and fibre optics have enabled the processing, storage, retrieving, and distribution of enormous amounts of information through communication networks.

The bandwidth explosion, as prophesied by Moore's law and Gilder's law[3], has resulted in significant reductions in cost and huge increases in speed and quantity. The same amount of information sent over a single cable within a second in 2001 is more than the monthly traffic over the entire Internet in 1997 (Gilder, 2000). It would cost $150,000 in 1970 to transmit a trillion bits of information from Boston to Los Angeles, but the price has fallen to about 12 cents nowadays. Sending a 40-page document form Chile to Kenya through an e-mail costs less than 10 cents, while faxing and express mailing take about $10 and $50 respectively (UNDP, 2001a).

The flexibility of Internet technologies and their compatibility with other wireless communication technologies[4] can help rural and

3 Moore's law refers to a statement by Intel co-founder Gordon Moore that said the industry should be able to double the number of transistors on a chip every 18 months or so. Gilder's law states that the amount of bandwidth doubles each 6 months, three times faster than the computing power.

4 Wireless technology works much the same way a mobile phone works: airwaves, instead of wires, are used to carry data from one point to another. A computer with a properly installed wireless network adapter can communicate with

remote areas bypass the construction of basic infrastructures and narrow the gaps in rural telecommunication coverage in developing countries. For instance, Wireless-in-Local-Loop (WLL) provides a more intelligent type of network with much lower cost. Depending on a cable network provided and maintained by local service providers, data and voice telephony can be transmitted over long distances wirelessly (Chapman & Slaymaker, 2002; Sood, 2003).

The Internet and its related technologies have fundamentally transformed the organization of work, the production mode, the market, the products and the consumption norm around the world. Its decentralizing architecture, easy access and affordability have provided wider opportunities for human development. Developing countries have been urged to harness the full potential of information revolution to alleviate poverty and seek sustained growth.

3.2 The Concept of Human Development

As mentioned in Chapter 2, the concept of development has undergone many transformations over the past few decades. Development was equated with economic growth and a higher standard living after World War II, measured and indicated by quantifiable yardsticks such as GNP or per capita income. This economic growth approach towards development lasted until the 1970s. With the rise of pluralist paradigm in development communication in the 1980s, development concerns started to shift to the non-material aspirations of people in the developing countries. The basic needs approach, also termed as "Another Development" concept, stressed the issues of social justice

---

another similarly configured computer or other wireless device, such as a printer with a wireless print adapter. Wireless technology provides speed and security comparable to a conventional wired network without the cables and their limitations.

and a balanced development without doing harm to the natural and cultural environment.

From the 1990s onward, human development and especially sustainable development, an all-encompassing concept, began to take over. According to UNDP, human development is about providing people with the choices to live life respectably and "creating an environment in which people can develop their full potential and lead productive, creative lives in accord with their needs and interests" (UNDP, 2001a, p.9). With achieving human freedom and well-being as the ultimate goal, human development means more than the accumulation of wealth and the rise of national incomes. Enlarging life choices is dependent on the enhancement of several basic human capabilities including long life expectancy, improved health condition and access to basic resources to ensure a decent standard of living and political participation (UNDP, 2001a).

With an aim of capturing a country's overall progress in achieving human development, Human Development Index (HDI) was first used in *Human Development Report 1990*. HDI focuses on three dimensions to make up its value: longevity, education and a decent standard of living, measured respectively by life expectancy, literacy rate, school enrolment and income per capita in purchasing power parity (PPP). Three supplementary indices have been developed to shed light on the distribution of the progress among different social segments: Human Poverty Index (HPI), Gender-related Development Index (GDI), and Gender Empowerment Measure (GEM) (UNDP, 2001a).

In order to highlight human deprivations in education, health and living standards, HPI-1 and HPI-2 have been created to reflect the separate conditions in developing countries and some selected Organization for Economic Cooperation and Development (OECD) countries.[5] Using the same HDI indicators, GDI is designed to reveal unbalanced achievement between men and women in the same three di-

5 HPI-2 focuses on deprivation in one additional dimension – social exclusion, measured by long-term unemployment rate (12 months or more) (UNDP, 2001a).

mensions of human development. Differing from GDI, GEM tracks the gender inequalities in the aspect of social and political participation and decision-making.

The developing world has witnessed impressive gains in all dimensions of development over the last 30 years. According to *Human Development Report 2001*, children today can live eight years longer by average than those born 30 years ago. The adult literacy rose from 47% (estimation) in 1970 to 73% in 1999. The rural families with access to clean water have grown five times. The average incomes in developing countries increased from $1,300 in 1975 to $2,500 in 1998 in real terms (1985 PPP US$). Despite the significant progress achieved, there are still large sections of population deprived of education, health and good living conditions. For example, the report also found that out of the 4.6 billion people in developing countries, 18.5% still remained illiterate, almost 21% lacked access to clean water, and around 52% were deprived of basic sanitation (UNDP, 2001a).

Confronted with gross inequalities in human development, 189 UN member states committed themselves to eradicating poverty, upholding human dignity and equality, and achieving peace, democracy and environmental sustainability at the historic UN Millennium Summit in September 2000. Eight MDGs were set to be achieved by 2015 with 1990 as the base year, including (a) eradicate extreme poverty and hunger; (b) achieve universal primary education; (c) achieve gender equality and empower women; (d) reduce child mortality; (e) improve maternal health; (f) combat HIV/AIDS, malaria and other diseases; (g) ensure environmental sustainability; (h) develop a global partnership for development. Under the eight goals, 18 specific targets are set out to ensure achieving the MDGs and 48 quantifiable indicators are formulated to monitor the progress within a specific country (UNDP, 2003).

There are strong linkages between human development indices and the MDGs. The four dimensions of human development captured in HDI – the issues of education, health, gender equality and a decent standard of living – are also reflected in the MDGs respectively, with the latter serving as a monitorable agenda for the global community.

The Millennium Declaration sees ICTs as an important tool with the potential to achieve MDGs. Target 18 of Goal 8 calls upon the UN member states to cooperate with the private sector to "make available the benefits of new technologies, especially information and communications" (UNDP, 2003, p.3). The corresponding indicators for the target are fixed telephone lines and cellular subscribers, personal computers in use and Internet users.

3.3 ICTs and Development: Optimists vs. Pessimists

There has been a wide range of opinions in the literature about the role of ICTs in promoting development even though research in this area has a short history. The impact of ICTs on social and economic development not only varies widely between industrialized and developing countries but also within the developing world. The question of whether the benefits outweigh the costs associated with ICT production and use has been hotly debated by scholars (UNDP, 2004b).

The advocates of ICTs argue that the technologies have the potential to contribute to human development by breaking barriers to human knowledge, participation and economic activity, which cannot possibly be achieved by other types of technologies. ICTs are becoming key carriers of information and knowledge for segments of population who have been deprived of the access to education. Distant learning based on ICTs, especially through radio and television broadcasting, offers remote villagers and the least privileged social segments wider access to education of various levels (UNDP, 2001a; UNDP, 2004a). The instantaneous transmission of information via ICTs and the interactivity involved in many of the multimedia education tools have largely changed the concept of teaching and learning and therefore enhanced the quality and efficiency of building and creating knowledge (UNDP, 2001a; UNDP, 2004a).

ICTs, often as an electronic and instantaneous conduit of opinion expression, greatly increase the opportunities of grassroots participation in public decision-making process. Governments around the world have employed ICTs as a means of making their operations more efficient and transparent as well as providing citizens with easy access to government services. The digitization of public service not only facilitates the information flow between citizens and their elected officials but also helps to establish a social environment of trust and reliability.

The most tangible contribution of ICTs towards economic growth is the generation of productivity gains as a sector of economic activity (UNDP, 2001a; UNDP, 2004a). Other than the productivity gains generated by ICT production sectors, the less visible, but the most profound long-term effects that ICTs have on economic growth result from ICT-using industries. In this respect, the benefits of ICTs are rather indirect and difficult to quantify because the technologies together with the information that they carry add value to the existing products and services. Mansell and Wehn (1998) argued that the information revolution triggered drastic changes not only in the creation and manufacture of ICT innovations but also in the traditional mode of production and organization of work.

ICTs also provide unprecedented opportunities for developing countries to adapt the structures of their economies and seek new models of development. The ICT sector in East Asia has witnessed rapid growth in recent years. ICT goods took up 28% of the region's manufacturing exports in 2003 and ICT production played a significant role in GDP rise in the region (Qiang, Pitt & Ayers, 2003).

Despite the high hopes pinned upon the new ICTs for the enhancement of human development, the notion of exclusion and divide is an important consideration in the literature where pessimistic sentiments about ICTs prevail. The most powerful argument about the downside of ICTs is that they have resulted in wider social and economic divergence not only among countries but across segments of populations within countries and such a divergence tends to be further exacerbated (UNDP, 2004b).

According to *the World Telecommunication Development Report 2003*, the penetration rates of fixed telephone lines and mobile phones in developed countries are more than five times those of developing countries. The report also found that developed countries[6] had 11 times more personal computer owners and eight times more Internet users than developing countries (ITU, 2003) (see Figure 1).

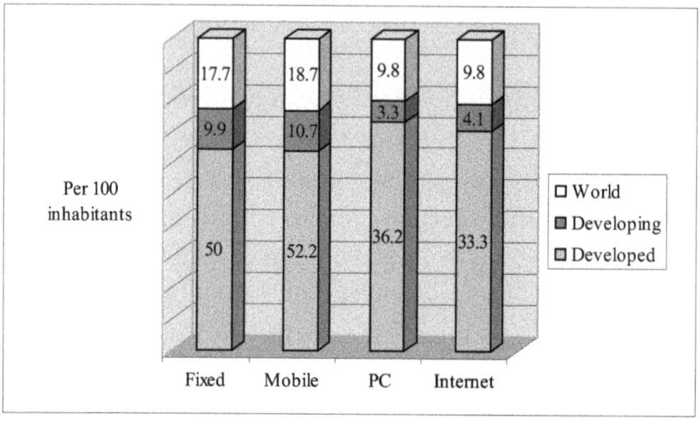

Figure 1. The digital divide between the developed and developing countries.[7]

It is argued that ICTs do not benefit the social sectors in developing countries the same way they do in industrialized countries. The pre-existing inequalities in access to natural and capital resources, social and cultural barriers to knowledge and information asymmetry make ICTs serve only to exacerbate the economic and educational marginalization. The introduction of a technology into a society tends to consolidate the positions of the privileged and make the rich richer and the poor poorer in terms of possession of both monetary wealth and information and knowledge (UNDP, 2004b).

There is a lot of scepticism with regard to whether ICTs can contribute to poverty reduction in the developing countries (Brown, 2001;

6 Developed countries include Western Europe, Australia, Canada, Japan, New Zealand and the United States (ITU, 2003).

7 Adapted from the World Telecommunication Development Report 2003 (p.77).

Heeks, 1999). Chowdhury (2000) contended that the poverty alleviation was a complex issue involving efficient production systems and physical infrastructures, to which ICTs seem to be of little avail. It is argued that " ICTs do not have any more to do with poverty and food security in the developing countries than rain dances have to do with rain" because high-speed Internet access cannot feed the poor (Chowdhury, 2000, p.3).

Furthermore, most of the research findings substantiating the ICT and economic growth nexus pertain to developed countries like U.S., Japan and several other OECD countries (Brynjolfsson & Hitt, 1996; Jalava & Pohjola, 2002; Jorgenson, 2001; Oliner & Sichel, 2000). There has been weak evidence to prove the positive links in developing countries because non-ICT sectors frequently generate more returns than the ICT sector. Kenny (2003) expressed doubt about the cost-effectiveness of ICT investment as far as the developing world was concerned. He pointed out that the present understanding of the possible relationships between ICTs and the achievement of social goals and economic growth was at a premature stage.

3.4 The Internet and Rural Development

3.4.1 The Potential of the Internet in Enhancing Rural Development

In view of the wide scope of the concept of human development in general and rural development in particular, this section discusses the connection between the Internet and five dimensions of development captured by the HDI in rural areas – livelihood, education, health, gender equality and participation.

a) The Internet and Livelihood

In addition to its impact on wealth creation and employment at the level of national economy, the Internet can also help the poor in fighting against poverty and hunger in direct and concrete ways (ITU, 2003; UNDP, 2004a). As agriculture is the chief source of income for people in many developing countries, the Internet can help farmers improve farming skills and agricultural practices through providing them with wider access to information on seeds selection, irrigation, temperature and humidity control, and fertilizers; and promoting information and knowledge sharing on farm innovations and the most recent advances in research and development. The rising crop and livestock yields as a result of scientific farming and increased earnings and savings from the use of online information can alleviate poor people of suffering malnutrition and starvation (UNDP, 2004a).

Another way that the Internet can help to increase incomes and savings of poor households is through online marketing of farm produce and purchasing farming facilities. In that aspect, the Internet has allowed farmers to see beyond their immediate milieu to look for business opportunities. It helps them get united as a group to participate in agricultural markets and control market changes by adjusting their production according to market demands and supplies (ITU, 2003).

The Gyandoot project in India established community-owned Internet kiosks in poverty-stricken rural areas. The agricultural database provided villagers with price information of prominent crops at the country's auction centres for a nominal charge. The prices paid to farmers increased 3% to 5% thanks to the cost saved from expensive and time-consuming trips to the market and the part of dealings taken away by middlemen and traders (Bhatnagar & Vyas, 2001).

b) The Internet and Education

The Internet cannot by itself achieve universal primary education, but it can be used as a powerful tool to enhance teaching and learning capabilities. One of the obstacles of getting all children to receive primary education lies in the lack of education infrastructures and teachers due to financial constraints. The Internet can bypass many of the limitations imposed by the traditional way of education (UNDP, 2004a).

In many developing countries, the majority of the rural population live in mountainous regions and remote islands with adverse geographical and transportation conditions. Distance education through online classrooms has made it possible to learn without the physical presence of teachers. Some pilot projects and software development efforts are just at their initial stages in rural areas. One notable example is an online education system designed by the Grameen Bank in India in collaboration with Nippon Electric Company (NEC) of Japan. The computer-based learning system is supposed to provide self-paced learning for rural families. The rural people can decide how much and what they want to learn through interaction with the system (Quibria & Tschang, 2001). China has used Modern Distance Learning Satellite Broadband Multimedia Transmission Platform to transmit video and multimedia channels to its western and other remote places since November 2000 (UNDP, 2004a).

The Internet can also offer training to the potential and actual primary school teachers. Many Internet-based teacher training programs help to alleviate the scarcity of skilled manpower in the field of education. In Latin America, 1,800 teachers from seven countries took part in a nine-month professional development course offered by education experts in Mexico. For six days every month, they received live broadcast of lectures from televisions at schools' computer labs and e-mailed their questions during the break of the broadcasts. The lecturers reviewed these questions together and answered them live. The participants commented that the distance classroom enabled them to take classes at their working places, whereas the enrolment into formal continuing education institutes required more time and money than they could possibly afford (Constance, 1999).

The web-based software and electronic learning materials can play a complementing role to the formal school teaching. The innovative use of the Internet as an information and knowledge delivery mechanism can help developing countries overcome the shortages of primary school textbooks. Malaysia has been studying the possibility of using electronic book appliances with textbook contents linked to the Internet to supplant traditional paper books (UNDP, 2004a). The Internet-based multimedia learning cultivates more autonomy and creativity among knowledge-seekers, turning learning experiences into a "self-assessed, self-paced and self-directed" experience (UNDP, 2004a, p.42).

Therefore, the Internet should be considered an alternative delivery system for developing countries to grapple with the problems of quantitative expansion of education for all social sectors. The Internet offers many possibilities of narrowing the pre-existing urban and rural disparities in terms of education access and quality.

c) *The Internet and Gender Empowerment*

The development of GDI and GEM as supplementary indices of HDI reflects the importance attached to gender empowerment in human development. A significant positive impact of the Internet on women's empowerment is offering women wider access to professional training, secondary education, on-the-job education and continuing education, and generating employment opportunities such as teleworking at homes.

The Internet has become an effective communication channel for rural women to receive information and knowledge and contact the outside world. In many developing countries, especially in the rural areas, women are still considered as an inferior and weaker gender under the influence of prevalent social values. They are deprived of equal opportunities for education, employment and political participation. The interactivity and networking characteristics of the Internet facilitate women activist groups to share information, coordinate actions and reach their target audience. Consequently, power synergies can be created to highlight gender issues and strive for equal rights for women. The Sustainable Development Networking Programme (SDNP) centres in Pakistan have been devoted to Internet skills training for women. The objective of the program is to help women get rid of their fear of technology and learn email and other basic computer skills so that the Internet can be put to family use. Up to mid-2000, around 760 rural women attended such workshops of short durations organized by the SDNP (UNDP, 2004a).

On the one hand, women can air their views on political issues and raise awareness about social injustices and obstacles in life over the Internet. On the other hand, female political candidates and activist groups can set up websites to reach large number of women and form or enrich their political groupings and alliances. Through more information and easier and flexible access to it enabled by the Internet, rural women are able to gain more understanding and awareness about their basic rights and hence use the Internet as a powerful tool to defend and promote their interests (UNDP, 2004a).

d) The Internet and Health

The Internet can have an impact on reducing infant and child mortality rates, improving maternal health, and halting and reversing the spread of HIV/AIDS, malaria, and other diseases in rural areas. One of the advantages of the Internet is that it offers at relatively low cost access to information and knowledge for health service professionals, the patients, caregivers and family members. It is estimated that around 60% of the births in the world are attended by midwives, nurses and doctors. Therefore, enhancing the training of health workers through diffusion of information about good maternal practice will not only reduce the mortality rates for both mothers and infants, but also improve their health tremendously (ITU, 2003).

Telemedicine is the most innovative use of new ICTs in remote healthcare. Through remote patient monitoring and diagnosis over multimedia communication links, both the quality and timeliness of the health service delivery can be assured. The Internet can help to connect hospitals at the national levels and those at the local levels so as to offer medical consultation over tele-radiology, tele-pathology and tele-cardiology to the rural population. Rural health professionals can also rely on database storage and information retrieval to improve the management of health information. The rural telemedicine centres in India, Apollo Telemedicine, provide medical services to around

50,000 rural people without access to proper healthcare. The Angel Programme in China has been set up as an ICT-based telemedicine network in the western regions to link to hospitals at the national level (UNDP, 2004a).

The Internet, together with the public broadcast media like radio and TV, can become effective means of raising awareness in the fight against HIV/AIDS, malaria and other diseases. It can help to disseminate public health messages and disease prevention techniques, and promote healthy lifestyles in the developing countries. Moreover, in many parts of the world, HIV/AIDS patients suffer from social discrimination and ostracism due to the public's lack of knowledge about the disease. The flexibility and anonymity nature of communication over the Internet can be used as a platform for those suffering from the disease to get counselling and comfort (ITU, 2003).

e) The Internet and Participation

As aforementioned, the introduction of E-government can render the government more efficient, transparent and accountable. Many developing countries such as China and India have begun using the Internet and computing systems to improve the efficiency of government services. The "Government Online Program" in China aimed to put 30%, 60% and 80% of government services online by 1998, 1999 and 2000 respectively. Several national projects have also been initiated by the Chinese government to facilitate transactions and payments to the government, tariff collection and tax filing (Quibria & Tschang, 2001). In India, the Gyandoot Project in Madhya Pradesh and Vision 2020 of the Chief Minister Chandrababu Naidu in Andhra Pradesh have provided rural people with access to information of government institutions, issuing copies of land records and online applications (Bhatnagar & Vyas, 2001).

E-government also facilitates citizens' participation in the decision-making process and functions as an effective means to check on corruption. The above-mentioned E-government services in India also provide feedback channels for people to file complaints. Rural people can go online to seek redress of grievances such as complaints concerning drinking water, quality of seeds and fertilizer, and functioning of village committees (Bhatnagar & Vyas, 2001). In one of the pilot ICT projects in suburban Beijing, the village committee uses the Internet to perform its daily administrative duties, greatly enhancing the communication between the governors and the governed. The committee regularly updates information regarding government grants received, public distributions, birth control policies, village election procedures and details about the candidates, making the governing process more transparent to the villagers (Hao & Zhao, 2004a).

3.4.2 Factors Affecting and Indicating Internet Diffusion and Usage

Despite the potential that the Internet has in combating rural poverty and fostering sustainable development, it requires an enabling environment of strong political commitment, liberalized telecommunication market, appropriate information infrastructures, participation of the private sectors and the NGOs, breakdown of social and cultural barriers, and capacity building (Chowdhury, 2000;UNDP, 2004a).

In recent years several E-readiness assessments have been developed to capture the development and divide in the ICT sector. According to the Centre for International Development at Harvard University, E-readiness refers to the capacity of a community or a society to participate in ICT development and its potential to exploit the technological innovations (UNDP, 2004a). Three major benchmarks of E-readiness have been constructed to offer a systematic comparison of the ICT development of countries around the globe – the Technology Achievement Index (TAI), the Networked Readiness Index (NRI), and the Digital Access Index (DAI) (UNDP, 2001a; ITU, 2003; WEF, 2003).

TAI was first used in the *Human Development Report 2001* by UNDP. It aims to capture "how well a country is creating and diffusing technology and building a human skill base – reflecting capacity to participate in the technological innovations of the network age" (UNDP, 2001a, p.46). Four components have been constructed to make up the TAI value: creation of technology, diffusion of recent innovations, diffusion of old innovations, and building of human skills. DAI is constructed by the International Telecommunication Union to measure a country's capacity to access ICTs, focusing on five factors affecting ICT access – infrastructure, affordability, knowledge, quality and usage.

NRI is another E-readiness measurement created recently to reflect countries' capacities to harness the opportunities offered by ICTs and detect factors that contribute to the building of those capacities. Networked Readiness is defined as "a nation's or community's degree of preparation to participate in and benefit from information and communication technology (ICT) developments" (WEF, 2003, p.3).

The Networked Readiness perspective suggests that a top country should be one with the highly developed ICT networks and the greatest potential to make use of the capacities of those networks. Three component indices have been constructed to capture the relationship: the Environment component index, the Readiness component index and the Usage component index (WEF, 2003).

The Environment component index is designed to measure how conducive the environment of a country is to the development and use of ICTs. It is made up of three subindexes: market, regulation and infrastructure. The market environment subindex reflects the preconditions of human resources and businesses to support a knowledge-based society, which is indicated by availability of scientists and engineers, quality of scientific research institutions, ICT manufactured exports and ICT service exports, etc. The political/regulatory index is constructed to reflect the effectiveness of the policies and laws governing the development and use of ICTs. Its constituent variables include quality of the legal system, laws relating to ICTs, competition in the ISP sector and foreign ownership restrictions, etc. The third subindex refers to the assessment of availability and quality of the access infrastructure for ICTs within a specific country. The overall infrastructure quality, waiting time for telephone lines, and the number of telephone mainlines, public pay phones and Internet servers are used to make up the value of the infrastructure index (WEF, 2003).

The Readiness component index focuses on the capacities of governments, businesses and citizens to exploit the opportunities offered by ICTs. Three subindexes compose the value of the Readiness index: individual readiness, business readiness, and government readiness. The individual readiness measures the preparation of a nation's citizens to utilize and benefit from ICTs. The variables that are used to measure this include adult literacy, the ownership of radios and televisions, and the availability and affordability of Internet access for individual users. The business and government readiness indexes are defined as the readiness of businesses and governments to employ ICTs. The business readiness subindex focuses not only on large corporations but also on medium and small-sized businesses and their willingness to utilize ICTs and make investment in elevating the ICT skills of their employees. The government readiness is used to reflect to what extent a government makes ICT development a priority of a nation's development initiatives. It stresses the policies and actions taken by governments to promote ICTs in the country in general and in the government services in particular (WEF, 2003).

The Usage component index is used to capture the degree and extent of usage of ICTs by the three stakeholders involved in the NRI framework. The individual usage, business usage and government usage constitute the value of the Usage component index. The individual usage is indicated by the number of personal computers, ISDN subscribers, cable television subscribers and the Internet users. The business usage is defined by variables such as computers installed in businesses, firm-level technology absorption and prevalence of foreign technology licensing. The presence of government online services and the volume of transactions that businesses have with governments are used to measure the usage of ICTs in the government sector (WEF, 2003).

Each of the three E-readiness assessment frameworks, TAI, NRI and DAI, distinguishes itself from others in terms of reflection and measurement of a country's readiness in participating in ICT development. TAI focuses more on the technology creation and diffusion and human skills building. The uniqueness of NRI lies in its identification and highlighting of three important stakeholders of ICT deployment: governments, businesses and individuals. Though DAI appears to be the most simplified index among the three, it incorporates the affordability and quality issues regarding the ICT access.

Though the three major international E-readiness assessment indexes have used varying composite of indicators to measure a country's readiness and capacity to exploit the opportunities offered by ICTs, the rationale supporting each framework is actually the same in nature. The diffusion and usage of ICTs in general and the Internet in particular are shaped by the existing social, political, and technological environments of a country. The ICT-related policies and regulations, government initiatives, telecommunication infrastructures, market environment and citizens' knowledge level and skill training are both important agents that condition and affect the employment of the Internet for achieving development goals and effective indicators of countries' strength and weakness regarding their participation in the networked age, which in turn serve as benchmarks for setting future development targets.

3.4.3 Issues of Concern

There is a paucity of empirical data to test relationship between the Internet and rural development (Quibria & Tschang, 2001). One possible explanation is that the technology has only started to diffuse in many developing countries over the last five years and it is still too early to gauge its social and economic impact. Owing to much of the hype about the role of ICTs in poverty reduction and education promotion, many ICT-related research and discussions do not seem to differentiate between what the Internet can be expected to achieve and what has actually been achieved. Consequently, there is a huge divergence between romanticized arguments of ICTs and empirical findings so far underpinning these claims. Furthermore, most of the empirical experiences are anecdotal in nature and lack theoretical conceptualization and sound methodological design. Despite shortage of theory-oriented research, several salient issues of concern arise in the extant case studies conducted in developing countries.

a) Policy Environment and Political Commitment

Though the advances of ICTs have made the information access and sharing much faster, cheaper and easier than before and technologies per se promise the possibility of a networked age where everyone is more or less connected, it is vitally important to take note of the underlying constraints of ICTs, especially concerning the integration of rural areas to the global knowledge economy (World Bank, 1995). Chapman and Slaymaker (2002) contended that as the reliance on market forces to ensure development and alleviate poverty proved to be ineffective, societies required government policies and initiatives to exercise social regulation and guarantee a balanced distribution of the benefits of ICTs among different social segments. The new landscape requires shifts in public policy both at the national and global levels to harness the technological transformations for public good.

Camacho (2001) asserted that the digital gap resulted from social and economic gaps and therefore the disparities would continue to widen with the uneven diffusion and adoption of the Internet. The author further stressed that the level of technology access and use was directly related to the electrification, telecommunication and education. Therefore the responsibility rested upon the government and its institutions to adapt its development strategies to the information needs of rural people.

The United Nations Commission on Science and Technology for Development (UNCSTD) Working Group on IT and Development concluded that every country should establish and implement a national ICT strategy to build capacities for producing, accessing and using ICTs. In order to be part of the global information infrastructure and the global knowledge economy, developing countries need to coordinate actions through national ICT strategy to "build the necessary scientific, technical, and engineering knowledge as well as the management techniques and social and economic institutions that are consistent with creatively using ICTs to reap the potential social and economic benefits" (Mansell & Wehn, 1998, p.7).

b) *Collective Access Points*

Based on the case studies of rural telecentres in Africa, Benjamin (2001) and Butcher (1998) argued that governments should take deliberate measures to ensure rural population's access to the Internet and community centres or infocommunes were considered an important implementation strategy to prevent marginalized communities from being further marginalized in terms of ICT access and utilization. This view is consistent with Mansell and Wehn's (1998) argument that poor countries should not emulate the predominant model of "one person-one telephone-one Internet access point" in the West because of large sized population and poor infrastructural base of ICTs.

The collective access points have been termed as telecentres, knowledge centres, information centres and information kiosks. Sood (2003) defined information kiosks as "multifunctional and networked installations at rural locations in developing environments that seek to use information and communication technologies (ICTs) to bring new access to content and services to rural citizen-consumers" (p.6).

India is way ahead of other developing nations with regard to its rural networking. It is documented that India houses over half of the information kiosk initiatives in the world with the rest scattered in other developing regions like Africa, Latin American and Southeast Asia and as many as 5.7 million rural users have actually benefited from ICT-enabled projects in India (Sood, 2003). Based on the successful experiences of the information kiosks in rural India, Sood refuted Bill Gates' statement that the Internet would not benefit poor people earning less than a dollar a day and argued that ICTs could help to ameliorate social and economic disparity between urban and rural areas. Sood (2002) contended that community or collective access to the network technology actually distributed the hardware and connectivity cost across many rural users, and therefore tremendously reduced the per capita cost of Internet access.

The rural information networks in India have enabled poverty-stricken people to seek information on medical care, make online transactions and send or receive emergency messages. The village information kiosks are also beginning to function as virtual institutions such as banks, postal offices, medical centres and public libraries through providing remote access to institutions at the municipal level, which saves huge possible costs in building institutional infrastructure catering to a limited population dispersed across large areas. This may further transform the traditional conceptualization about rural communities (Sood, 2003).

Best and Maclay (2002) proposed that networking the rural regions would benefit from the "Metcalfe Effect", which meant that the value of the network to other Internet users grew with each additional user. In other words, the more users get connected with the network, the more value the Internet will be endowed with. The linkages among the diverse transaction parties over long distances will enhance the value of the whole rural network. Another advantage of the Internet kiosks lies in that they can ensure equality in terms of content and service access and help to improve the quality of government services and increase transparencies and accountability (Sood, 2003).

c) Sustainability

As most of the rural information centres are pilot or demonstration projects financially supported by governments or external donors, the sustainability of the Internet services has become a big concern. There is a lack of understanding about the full costs involved in running information centres as well as the social and economic benefits that accrue from improved access to information via the Internet (Quibria & Tschang, 2001).

Several large-scale projects in India have proved to be financially viable under the joint sponsorship of the local government and private sectors. The state governments of Andhra Pradesh and Karnataka have cooperated with the private telephone companies (such as Reliance) to install fibre optic infrastructure and information kiosks in villages. Around 7,500 information kiosks have been set up in each of the two states, which cover one fourth of the total villages. Each user is charged a transaction fee for sending application forms of obtaining income/caste/domicile certificates, accessing market information and putting one's commodities for sale online, etc. The service-fee-charging model proves to be an effective way of keeping the rural Internet services financially viable, sustainable or even profitable (Quibria & Tschang, 2001).

However, the public-private partnership that has functioned in India does not seem to work in rural China. As Ulrich (2003) pointed out, it was not possible for most villages to support viable telecommunications businesses based on the relatively small population sizes. Without subsidies to villages for paying online access charges, there was little incentive for them to expand the usage. As a result, charging for services for future sustainability was not the answer to the expansion of ICTs in China, at least at the present stage. Other factors which may influence the usage and viability of Internet services include the relevance of Internet content to the needs of farmers, the farmers' command of basic IT skills and the complimentary role of the Internet to traditional means of communication (Quibria & Tschang, 2001).

Despite some heartening experiences from developing countries, it still remains unclear whether the pilot projects can be replicated beyond the demonstration phase and broadened in scope. Much more micro-level research is needed to understand and assess the impacts of ICTs on the livelihood of the rural people, such as to what extent these rural Internet-based services can be accessed by the rural poor and how they help to generate incomes, promote education and health, and enhance participation.

3.5 Summary

The rapid development and spread of new communication technologies has generated a new wave of enthusiasm among developing countries to harness these advanced technologies for accessing the global information infrastructure and participating in the knowledge economy. Concomitantly, development at the same period starts to evolve into an all-encompassing concept, which stresses the enhancement of basic human capacities including a decent standard of living, improved health, education and participation.

The architecture and technological characteristics of the Internet have promised wider opportunities of alleviating poverty, enhancing education, promoting gender equality, improving health and encouraging participation for rural people of the developing countries. However, the striking reality is the contradiction between the potential for ICTs to address development concerns and the widening divergence between the developed and developing worlds and between the urban and rural areas within countries in terms of the diffusion of ICTs as well as the distribution of their benefits.

The same ambivalent attitude is predominant in the literature regarding ICTs and rural development. The optimists including international organizations such as UNDP, World Bank and ITU urge developing countries to make huge investment in their national information infrastructures so that their existing development problems can be tackled with the aid of ICTs. One of the conclusions made by the UNCSTD working group is that "Although the costs of using ICTs to build national information infrastructures which can contribute to innovative 'knowledge societies' are high, the costs of not doing so are likely to be much higher" (Mansell & Wehn, 1998, p.7).

However, the voices of pessimists are equally vociferous. They argue that the research findings validating the ICT and economic growth nexus only pertain to developed countries and there is a shortage of clear evidence to prove the positive links in developing countries. Anderson, Crowder and Dion (1999) noted that getting wired was a means rather than an end and the utilization of the Internet deserved more attention than mere access. The authors warned that developing countries did not seem to draw lessons from disappointing experiences in the past regarding technological transfer. More emphasis should be put on people, organizations and process instead of the technologies per se. The pessimists hold that money in developing countries should be better spent on elementary education and basic infrastructures such as transportation.

The conflicting views regarding ICTs and rural development are also considered part of the academic debate on the relationship between society and technology. Technological determinism, effectively the opposite of social determinism, views technology as a key governing force in society. Chandler (1995, Technological autonomy section, para.1) pointed out "Rather than as a product of society and an integral part of it, technology is presented as an independent, self-controlling, self-determining, self-generating, self-propelling, self-perpetuating and self-expanding force". With an increasing recognition in recent years, the social shaping of technology argument underscores "the 'choices' (though not necessarily conscious choices) inherent in both the design of individual artefacts and systems, and in the direction or trajectory of innovation programmes" (Williams & Edge, 1996, p.857). The central issue here is whether technological development is primarily determined by its social context, or whether technology determines the social context including its norms and values.

Before the Internet can make a difference on rural development, the technology needs to be diffused and adopted by the rural people. As the implications that the Internet has on rural development are in effect imbedded in the diffusion and adoption of the technology among rural people, Chapter 4 reviews the major concepts of the diffusion of innovations theory, the research findings of the available studies as well as the dominant research orientation.

4. Diffusion of Innovations

With its roots in rural sociology, the diffusion of innovations model has been formulated on the basis of the research findings of agricultural extension studies. Agricultural extension is defined as "a professional communication intervention deployed by an institution to induce change in voluntary behaviours with a presumed public or collective utility" (Röling, 1988, p.49). Whereas agricultural extensions deal with how political, socio-economic, cultural, technical, and human behaviour factors influence agricultural change, the diffusion of innovations model extends the research paradigm to other disciplines such as education, public health, marketing and communication (Delman, 1993).

A seminal study conducted by Ryan and Gross (1943) on the diffusion and adoption of hybrid corn among Iowa farmers laid the foundation for the methodology as well as theoretical framework of later diffusion studies. Everett M. Rogers formalized the diffusion of innovations theory by summarizing the extant diffusion studies in his *Diffusion of Innovations*.

As the basic concepts of the diffusion of innovations theory will be applied throughout this study to explicate the diffusion and adoption of the Internet among Chinese farmers and its implications on rural development, this chapter summarizes the main elements involved in the diffusion studies as well as the findings so far in order to derive lessons learned from past research. Attempts will also be made to critique the weaknesses as well as biases imbedded in the diffusion model, which may help to set the departure points for the present study.

4.1 The Main Elements

Rogers (1983) defined the diffusion as "… the process by which an innovation is communicated through certain channels over time among the members of a social system" (p.5). Therefore, the four main elements involved in the diffusion process are the innovation, communication channels, time and the social system.

Rogers (1983) conceptualized an innovation as "an idea, practice, or object that is perceived as new by an individual or other unit of adoption" (p.11). Rogers held that technological innovations brought both uncertainty and reduced uncertainty to the possible adopters. Reduced uncertainty meant that technologies offered certain advantages that helped individuals in meeting their needs and solving their problems. Driven by such incentives, the possible adopters engaged themselves in information seeking and processing activities to reduce their uncertainty about the expected consequences of technological innovations. The decision as to whether to accept or reject an innovation was made when an individual's feeling of uncertainty decreased to a "secure" level. After the adoption of the innovation, the adopters would seek further evaluative information and the uncertainty was in turn further reduced.

Five major characteristics of innovations offered explanations why certain innovations spread rapidly and others slowly within a social system. They were relative advantage, compatibility, complexity, trialability and observability. Rogers (1983) proposed that the more advantages that a particular innovation was perceived to have than the previous one; the more rapidly it would be adopted. The advantages were not confined to an increase in wealth; they included satisfaction of needs and provision of convenience, etc. Moreover, an innovation should not go against the pre-existing value and belief systems of the potential adopters and must have the efficacy to satisfy their concrete needs.

Complexity meant that innovations that did not require grasp of new skills and knowledge would be adopted more rapidly than those that were beyond the cognitive level of the potential adopters. In addition, if an innovation could be tried on a limited basis, the experimentation helped to reduce the uncertainty that individuals had towards the innovation and could speed up its rate of adoption. Observability suggested that if the advantages of an innovation were easily visible among the peers of those who had actually adopted the innovation, the diffusion could be activated much more quickly (Rogers, 1983).

Diffusion was considered a special type of communication process as it dealt with novel ideas that brought about social changes. The diffusion communication was seen as a two-way process of convergence involving creating and sharing of information with an aim to achieve mutual understanding. Rogers and Svenning's (1969) studies of modernization among Colombian peasants employed the S-M-C-R-E model to illustrate the role of communication in the adoption of new ideas or behaviours (see Figure 2).

Rogers (1983) held that mass media channels were the most efficient means of generating awareness knowledge about an innovation among the potential adopters, while interpersonal channels were more potent in persuading individuals to accept or reject an innovation via face-to-face exchange of information. The results of diffusion studies showed that the majority of individuals did not refer to scientific findings to make a judgment of innovations. Instead, their evaluation was by and large influenced by the views of those adopters whom they knew. Rogers (1983) argued that "this dependence on the communicated experience of near-peers suggests that the heart of the diffusion process is the modelling and imitation by potential adopters of their network partners who have adopted previously" (p.18).

The dimension of time was taken as an important variable in the diffusion research. The time sequence order in the innovation decision process, the relative earliness and lateness of adoption among the participants and the group adoption rate within a social system were all concerned with the measurement of the time dimension.

Rogers (1983) conceptualized that five steps were involved in the innovation decision-making process as far as individuals were concerned – knowledge, persuasion, decision, implementation and confirmation. The knowledge step took place when an individual first became aware of the functions of an innovation. It was followed by the persuasion step when the individual developed a kind of attitude either favourable or unfavourable towards the innovation. Then it was time for the individual to arrive at a decision as to accepting or rejecting the innovation. Implementation meant that the individual put the technological innovations into practice. Confirmation occurred when the individual either reinforced his original decision or went against it based on whether the implementation experiences were satisfactory or not.

Source →	**Message** →	**Channel** →	**Receiver** →	**Effect**
Scientists, government planners, change agents, other originator of new ideas	New methods for social functioning: technological, economic, political or social	Mass media and interpersonal media	People of less developed nations	Change in knowledge, attitudes and overt behaviour of the receivers

Figure 2. Diffusion perceived as a communication process.

The members in a social system were classified into five categories on the basis of their innovativeness as determined by the time period for an innovation to be adopted. They were innovators, early adopters, early majority, late majority and laggards. Innovativeness meant "the degree to which an individual or other unit of adoption is relatively earlier in adopting new ideas than the other members of a system" (Rogers, 1983, p.22). Diffusion researchers had used innovativeness as an important yardstick to measure the success of development projects because innovativeness signified changes at the behavioural level rather than at the cognitive and attitudinal level.

Adoption rate examined the time that an innovation took to spread within members of a social system. As far as the rate of adoption was concerned, innovations in a social system were used as units of analysis rather than individuals. Rogers (1983) identified that the diffusion pattern of an innovation followed an s-shaped curve when the cumulative frequencies of adoptions were measured over the lapse of time. The relative rate of adoption directly determined the steepness of the curve and it itself was subject to the influences of many factors. If an innovation was perceived to be advantageous and compatible with the pre-existing social values of the potential adopters, then its adoption rate could be more rapid as compared to other innovations that were lacking in those attributes. Rogers contended that the adoption rate of the same innovation in different social systems could be in sharp contrast with each other. Consequently, a social system could affect the diffusion of innovations directly through its various structures and indirectly through the individual members within it. The relative adoption rate of innovations had been one of the research foci of the innovation studies.

As the diffusion process took place within a social system, it was therefore under the influence of its social structures. In fact, the system as well as the characteristics of its individual members exerted influence on the diffusion and adoption of innovations. The individual innovativeness was usually determined by characteristics of the individual as well as the social structures within which he or she lived (Rogers, 1983) (see Figure 3).

Besides the influences of social structures, social norms, opinion leadership and change agents could also have a strong effect on the diffusion and adoption of innovations. Opinion leaders, typified by certain characteristics differentiating them from their followers, tended to have higher socioeconomic status, higher level of conformity with the social norms, and a wider contact with the external communication. An opinion leader often held an influential position in the communication structure of a social system which enabled him to "serve as a social model whose innovative behaviour is imitated by many other members of the system" (Rogers, 1983, p.28).

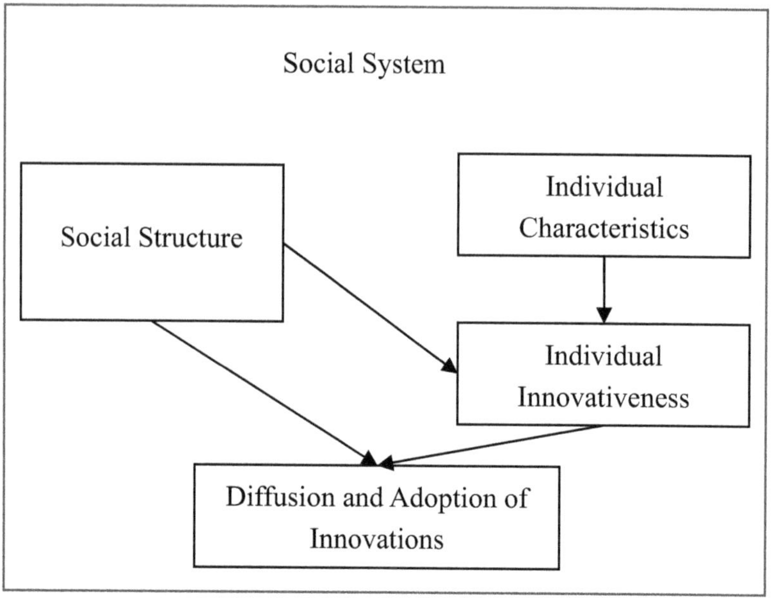

Figure 3. The diffusion and adoption of innovations within a social system.

Rogers (1983) defined a change agent as "an individual who influences clients' innovation decisions in a direction deemed desirable by a change agency" (p.28). As change agents were often experts with high professional training and a good command of technical knowledge, they often met difficulties in disseminating the innovative knowledge since they did not share much in common with their clients. Consequently, they had to depend on the influence of opinion leaders or the efforts of the local aids to facilitate the diffusion process.

Another way that a social system exerted its influences upon the diffusion process was through moulding innovation decisions. Rogers (1983) identified four types of innovation decisions: optional, collective, authoritative and contingent innovation decisions. Optional innovation-decision left the power of decision as to accepting or rejecting innovations with individuals regardless of other members' decisions. With collective innovation-decision, the adoption of innovations was agreed upon by the majority of a social system.

Authoritative innovation-decision meant that the social or technical authorities in a system made choices regarding the adoption or rejection of an innovation, which left little leeway for individuals to go against the "orders". With the last type of innovation-decision, the adoption decision of one innovation was dependent upon a previous innovation decision. Sometimes a prior innovation must be endorsed first so that an individual's optional decision could be made possible.

4.2 The Main Propositions

The diffusion model takes the adoption rate as both the ultimate goal of the diffusion activities and the essential yardstick to measure their effectiveness. In order for an innovation to rapidly spread among members of a social system, Rogers and Scott (1997) argued that most efforts should be spent in pushing the adoption of the innovation to the point of critical mass, where the further adoption rate started to be self-sustaining regardless of the diffusion efforts. With the s-shaped curve, the area between 10% and 20–25% signified the time period when the diffusion started to take off. This period was taken as the significant part of the diffusion process in that the subjective judgment of an innovation began to circulate among peers via interpersonal networks in a social system.

Therefore, the diffusion activities should be concentrated on the 13.5% early adopters who tended to play a decisive role in the successful diffusion of an innovation. The positions of the early adopters as opinion leaders enabled them to exert great influence on the potential adopters as they enjoyed great respect and remained central in the interpersonal networks. Through their interpersonal connections, the early adopters spread their opinions about innovations and their very action of adopting an innovation and subjective evaluation about it would help in reducing the degree of uncertainty that peers around him had (Rogers, 1983). Rogers and Scott (1997, p.10) suggested that the extension efforts should be targeted at the early adopters or the opinion leaders so as to realize "the potential power of the diffusion process to multiply existing resources devoted to outreach".

Rogers (1983) held that the two-step flow model of mass communication should be employed to facilitate the diffusion of new ideas within a social system. The innovative ideas from a source often reached the opinion leaders first via the mass media and then were passed onto their followers. The integration of mass media with interpersonal channels in information sharing could lead to a better mutual understanding between the change agencies and their intended audience.

Rogers and Svenning's (1969) study concluded that localite/cosmopolite communication channels did a better job than interpersonal/mass media categorization in explaining the individual decision-making process as far as the Colombian peasants were concerned. Localite channels had their origin within receivers' social system while cosmopolite channels originated from outside (see Table 1).

The research findings showed that mass media and interpersonal channels tended to exert varying influences on the different stages in the individual decision-making process. Mass media and cosmopolite channels were more influential at the awareness stage and trial stage, whereas interpersonal and localite channels seemed to be more effective in helping farmers develop an interest and arrive at decisions as to innovation adoption or rejection. This was mainly due to the fact that extension agents and farm-supply personnel (cosmopolite) helped to introduce an innovation to a village (awareness stage) and provided instructions on how to use it at the trial stage. However, opinion leaders and early adopters tended to have a greater influence in helping the potential adopters foster attitudes or make decisions.

Table 1
Categorization of Illustrative Channels as Localite/Cosmopolite and Interpersonal/Mass Media

Nature of Channel	Place of Origin	
	Localite	Cosmopolite
Interpersonal	Neighbour	Extension agent
	Village Council	Wandering storytellers
	Relative	Salesman
Mass media	Village newspaper	Radio
	Wall posters	Television
		Cinema
		City newspaper

Note. From *Modernization among peasants: The Impact of Communication* (p.128), by E.M. Rogers & L. Svenning, 1969, US: Holt, Rinehart and Winston.

4.3 Criticisms of the Diffusion Approach

The diffusion of innovations model has been criticized for both its methodological and conceptual weaknesses. The criticisms revolve around the innovation bias, the flaw in survey design, lack of concern for the socioeconomic consequences of innovation diffusion, and inattention to message content.

First, one of the implicit assumptions of the diffusion model is that the innovations will benefit groups or individuals who adopt them. As the adoption of innovations tends to have both desirable and undesirable consequences for the farming households and the rural communities, the risks innovations involve have not been taken into serious account by the diffusion studies (Van den Ban & Hawkins, 1988).

The diffusion researchers more often than not attempt to introduce Western agricultural practices to rural communities engaged in century-old mixed-cropping farming. A lot of cases in the developing countries suggested the incongruence of advanced innovations with the traditional subsistence farming prevalent in the local communities (Melkote & Steeves, 2001). Whyte (1991) and Horton (1991) illustrated the creativity and rationality of small farmers in Latin America, whose cultivation and seeding methods were superior to the innovative technologies typical of Western farming. The ill adaptation of innovations to the local situations leads to "the painful contradiction between diffusion theory and its practice" (Melkote & Steeves, 2001, p.227). The belief in the one way flow of information from exogenous sources to rural communities views traditional farming as primitive and uneconomical and forecloses the possibility of reverse flow of ideas and practices from farmers to administrators and scientists (Melkote & Steeves, 2001).

Secondly, the dominance of one-shot survey in the diffusion studies goes against the conceptualization of communication as a process. The survey can only yield recall data about past adoption behaviours collected at a single point in time and fails to predict the trend of adoption and consider the existence of other viable alternatives. Another weakness with the survey method lies in its inability in controlling the time-order sequence of independent and dependent variables. Rogers (1976) argued that in the survey methods, the dependent variable was measured based on the recall data on the past adoption behaviour, whereas the independent variable was measured according to the present tense. This leads to the inversion of occurrence between the two variables in time sequence, which renders the research design questionable.

The quantitative measure of the media exposure such as receivers' self-report of the degree of exposure to the print media, radio broadcast and movie watching fails to capture their preferences of media content, disparities in media use and perception among individuals or groups, and the quality and relevance of the programmes (whether pro-development or not) (Golding, 1974).

There is also a lack of concern for the cognitive dimension of communication effects, like the internalization of the message content among receivers in the diffusion of innovations studies. The measurement of innovation adoption as a dependent variable stops at the awareness level, i.e. whether the name of the innovations has been heard of and the actual command of knowledge tends to be neglected. Rogers and Shoemaker (1971) made attempts to refine their earlier conceptualizations and differentiated the knowledge acquired through the diffusion process into three categories: (a) awareness-knowledge (the individual was aware that certain innovations existed), (b) how-to knowledge (information about how to use an innovation properly) and (c) principles-knowledge (the basic facts and principles about the innovation). Shingi and Mody (1976) proposed that there was a need to take into account the depth and adequacy of information acquired by the receivers when farmers' information profiles were developed to predict innovative behaviours.

Thirdly, the innovative technologies tend to favour progressive farmers with higher socioeconomic status and easier information access and discriminate against resource-poor farmers due to their lack of financial resources and information access. Melkote and Steeves (2001) argued that, "the progressive farmer strategy coupled with the pro-innovation, pro-source, and pro-literacy biases worked against poorer farmers" (p.240). The imbalance leads to a vicious circle of development – the advanced farmers are more likely to be successful in their adoption of innovative technologies, get empowered through knowledge and financial returns and thereby increase their information seeking whereas resource-poor farmers are in a disadvantaged position in their fighting with social forces and become fatalistic and resistant to innovations, owing to their lack of control over the environment around him. Therefore, the diffusion of innovations may exacerbate the socioeconomic, gender and class inequalities, and the disparities in access to information even within the same rural region.

Last, the diffusion of innovations model lacks concern for the message content diffused through extension agents to farmers. This is largely due to the presumption that an innovation itself is considered the message. The limited number of extension agents has confined their contact to only a fraction of farmers, and the eager-information-seeking farmers are usually given undue importance in the information disseminating process (Roling, Ascroft, & Chege, 1976; McAnany, 1980). Consequently, information on innovations either cannot reach resource-poor farmers at all or gets distorted via the channel of word-of-mouth.

As the diffusion campaigns often take less heed of the sharp contrast among receivers in terms of their socioeconomic status and cognitive ability, diffusion strategies should be sequential and research-based in order to achieve the desired effect of communication. Messages should be well designed in accordance with the complexity of innovation, level of cognition of receivers and their information needs. It is recommended that the diffusion of agriculture innovations can be broken up into several stages to ease the difficulty of adopting complex innovations. The social marketing campaign approach is seen as an improvement on the diffusion model by audience segmentation and strategic planning (Melkote & Steeves, 2001).

4.4 A New Research Perspective

The weakness of the diffusion research has resulted in the preoccupation with individual innovativeness studies and a lack of understanding on the influences of social structures on the diffusion process, the social consequences brought by the innovation adoption, and the possible discontinuance or adaptation of innovations at the post-adoption stage (Rogers, 1983; Zhu & He, 2002).

Rogers (1983) pointed out that the consequences of innovations had been less researched and most of the existing studies stopped at the level of description and categorization rather than providing predictions about circumstances under which certain consequences would take place. The diffusion researchers tended to concentrate on the decision-making process in terms of innovation adoption and downplay the implementation and the consequences associated with it.

The research gaps have given rise to the shifting of focus in diffusion studies in recent years. As Rogers' seminal model of diffusion of innovations addresses a variety of innovations ranging from new farming practices to medical inventions, whether the extant conceptualization fits in with the extension of new communication technologies also poses new challenges to the classical diffusion model.

Under the influence of the pro-innovation bias, many efforts have been spent in investigating antecedent variables predicting innovativeness and the adoption of an innovation is taken as an ultimate end in the diffusion process. Zhu and He (2002) argued that the Internet diffusion study should include the post-adoption process as the skill-dependent technology required not only the initial investment in purchasing the hardware but also the technical competence of its users. They proposed that Internet adoption and use should be conceived as separate stages in the diffusion process and the uses and gratifications theory should be integrated with the diffusion of innovations model to shed more light on the continuous and changing process of the adoption and utilization of technologies.

Zhu and He's (2002) study on Internet adoption and use in China concluded that the adoption behaviours of the urban users were, to a large extent, determined by the perceived characteristics of the Internet (PCI, operationally defined as relative advantage, compatibility, ease of use, results demonstrability and image) and the perceived popularity of the Internet (PPI, the equivalent variable for measuring the contextual influences such as social norms), whereas the sustained use of the Internet depended solely on the perceived needs for the Internet (PNI, the needs that cannot be possibly achieved by competing media channels or content).

The identification of PCI, PPI, and PNI as decisive psychological factors predicting the adoption and use of the Internet has provided an alternative approach to examining and analyzing the diffusion process.

As a deliberate attempt to overcome the pro-innovation bias, it addresses the possible discontinued use or rejection of the Internet after the adopting behaviours, driven by less needs for the technology or less conveniences provided as compared to the competing media channels or content. Since media uses are primarily need-driven, the subjective perceptions and social pressure cease to play a part in sustaining the continuous use. The incorporation of uses and gratifications theory into the diffusion research enables the extension of the diffusion process, which no longer ends with the adoption. Instead, the diffusion model starts to be concerned with the active usage and the possible impacts innovations may have on the adopters or the society they live in.

The new research perspective suggests that more efforts should be made to discover how potential adopters perceive the innovations and what factors motivate their adopting behaviours and sustain the continued usage. This calls for a broader approach to the diffusion research, which centres on appreciating the possible consequences and implications associated with the innovation adoption from the users' point of view.

With regard to new communication technologies, innovation researchers often observe that even though these technologies may be widely acquired within firms, they are only sparsely used among individuals. Fichman and Kemerer (1999) argued that there might be a pronounced difference between acquisition and deployment of the technologies, which he termed as "assimilation gap" (p.255). Therefore, taking the cumulative purchases of a technology as the sole indicator of adoption may not present a true reflection of the diffusion process and invalid judgements could be made concerning the future prospects of the technology. In examining the diffusion of wireless Internet among employees in the workplace, Wei (2006) found that only about one-fifth of the respondents used Wi-Fi at the time of the study. His findings further substantiated the possibly widening gap between acquisition and usage. Similarly, Steinfield pointed it out that despite the building of new communication infrastructure within organizations, "individuals ultimately choose to use one medium or another to communicate with others" (2002, p. 249).

As the diffusion initiatives aim at voluntary behavioural changes of individuals, the perceptions held by the potential adopters about an innovation and their various needs such as information seeking and social networking should be carefully examined since they tend to be the key factors in motivating individuals to change and engage in self-renewing efforts to reinforce the changes (Delman, 1993). An increasing number of studies have been carried out to understand the information needs of the targeted audiences of diffusion initiatives (Bowden, Kromer, & Tobia, 1994; Mullaly-Quijas, Ward, & Woelfl, 1994; Wallingford et al., 1996). In their studies of diffusing information resources among health professionals, Wallingford et al. (1996) argued that a thorough understanding of the information needs of the intended audience was crucial to the effectiveness of outreach activities.

4.5 Summary

Most of the local level development communication planning and execution from the 1950s to the early 1970s were under the guidance of the diffusion of innovations model (Melkote & Steeves, 2001). The theory posits that increased development and modernization can be achieved through introducing technological and agricultural change into the developing countries and the adoption of innovations is supposed to result from the communication of innovation via media and change agents (Okigbo, 1985).

The diffusion of innovations research derived from the classic agricultural extension model and reflected the dominant trends of communication studies throughout the 1960s. The major concern at the time was the effects of a particular source, channel and message on changes in knowledge, attitudes and overt behaviours of receivers.

The focus on communication effects under the S-M-C-R-E model led to the emphasis on behavioural changes through increased mass media exposure.

The diffusion model has been criticized for its pro-innovation bias, which contributes to the concentration of research efforts on predicting the innovativeness of individuals and the decision-making process of innovation adoption. Consequently, the influences of social structures on the diffusion process, the social and economic impacts of innovations as well as the possible discontinuance and rejection taking place after the initial adoption remain largely understudied. In order to fill the gaps, a new research perspective starts to emphasize the perceptions held by the targeted users as well as the motivating factors driving their adoption and continuous usage of innovations. The following chapter will synthesize the theories, views and research findings covered in Chapter 2 to Chapter 4 and formulate the research questions that the present study sets out to address.

5. An Overview of Literature

5.1 The Evolution of the Concept of Development

Through over a half century's research on development communication, the concept of development has undergone many transformations. We have observed the evolution of the development concept from a sheer pursuit of quantitative accumulation of economic gains to the striving for a plurality of economic, moral and spiritual goals.

The modernization paradigm prescribed a unidirectional and irreversible model of development from an underdeveloped traditional society to an industrialized and materialistic modern society. In the 1960s, development was defined in quantitative terms measured by economic growth and the total volume of goods and services. In Rogers' 1969 conception, development meant the pursuit of higher economic growth and living standards through adopting modern production methods and social organization (Rogers & Svenning, 1969).

A decade later, with the maximization of goods and services incurring many negative social influences such as high level of unemployment and widened social inequality, the dependency school identified structural deficiencies as impediments to further development. As the structural impediments were seen as both external and internal, radical measures such as land reform and redistributing resources were carried out at the domestic level and more equitable trade relations were pursued at the international dimension (Jayaweera, 1987). The central argument of the dependency school was that the underdevelopment of periphery countries was directly caused by the development of core countries and therefore the solution to underdevelopment was to disassociate themselves from industrialized countries and seek self-reliance. There was little primary difference between the modernization and dependency paradigms as far as the content of development was concerned. Both schools emphasized economic growth as a yardstick for development.

Starting from the mid-1970s, the concept of development began to expand as many drawbacks of industrialization and urbanization such as severe depletion of resources and environmental pollution were recognized. Rogers' (1976) definition captured much of the sentiment and took both social and material development into consideration. Development was interpreted as:

> A widely participatory process of social change in a society, intended to bring about both social and material advancement (including greater equality, freedom, and other valued qualities) for the majority of the people through their gaining greater control over their environment (Rogers, 1976, p.225).

Many policies in the 1970s were targeted at political and socioeconomic equity through redistributing the economic surplus. The incorporation of education, literacy, health, infant mortality and life expectancy into the life quality indicators suggested dramatic improvement in the conceptualization of development.

By the 1980s, the developing countries realized that they still could not achieve the same level of development as the industrialized countries even if they had been engaged in structural reforms and endeavours of securing equitable distribution of social benefits. The concept of "Another Development" was raised by the pluralist school who advocated a balanced development highlighting human rights, equal access, participation, democracy and environmental protection. The pluralist perspective promotes self-reliance and decentralization of development based on local initiatives and grassroots participation.

The "Another Development" or "basic needs" approach was a step closer to the crystallization of the goal of human development. Since the 1990s, international organizations such as the UNDP and Food and Agriculture Organization of the United Nations (FAO) have helped to sharpen the focus of development and regarded enlarging life choices through the enhancement of several basic human capabilities and developing human potentials as the ultimate goals of development. The basic human capabilities involve a decent standard of living and improved health, education and participation. The human development concept is adopted by the study to examine the implications of Internet diffusion and usage on five dimensions of the life of rural people: livelihood, education, health, gender equality and political participation.

5.2 The Role of Communication in Development

The understanding of the role of communication in general and mass media technologies in particular in promoting development does not witness much improvement over the last five decades, even with the unfolding of the information revolution and the ensuing rapid diffusion of new ICTs in developing countries. High hopes have been pinned upon media technologies to bring about desirable social, economic and political changes despite a paucity of research findings to empirically verify the contribution of ICTs to achieving the development goals in developing countries.

The three paradigms in development communication accorded communication different roles in promoting development. Under the modernization paradigm, communication was an essential element for developing countries to move from a traditional to a modern society. Mass media were viewed as most effective in fostering a pro-development environment and inculcating modern attitudes and psychic aptitudes among individuals, whereas interpersonal communication was more potent in bringing about attitudinal changes and altering values and group practices. The communication process followed a top-down model with one way flow of information from senders to receivers and both development and communication strategies were characterized by centralized planning.

The perception of the role of communication in development in the 1950s and 1960s was under heavy influence of the research findings on media effects because development communication was specifically concerned with the impacts that media could bring about upon development. In Lerner's (1958) model of modernization, media exposure was considered both a vital phase of the modernization process and an index of a modern society. Rogers and Svenning (1969) argued that the transition from a traditional to a modern life style was based on communication and adoption of new ideas. As mass media were believed to carry elements of modernity, developing countries were encouraged to establish mass media channels and make heavy investment to improve their media infrastructures. In 1961, the UNESCO recommended the minimum number of mass media for developing countries – at least 10 copies of daily newspapers, five radio sets, two cinema seats and two television sets should be provided for per hundred inhabitants in each developing country (Yu, 1977).

As an antithesis to the modernization paradigm, the dependency paradigm focused on unbalanced information exchange at the international level. It accused the Western media of using liberalism as a tool for economic and ideological expansion. With the UNESCO debate over a new global economic and information order, the developing countries were called upon to sever up their mass media with the unbalanced information and communication system and reject the Western beliefs and values (Stevenson, 1988). Therefore, the political and ideological role of communication was more emphasized under the dependency paradigm. However, the dependency school's radical views towards solutions to underdevelopment did not result in the desired democratization of communication in the developing countries, but an increase in state intervention and tight regulation in the production, distribution and operation of the mass media (Servaes, 1991).

The inapplicability of the modernization and dependency paradigms to the developing countries and the liberation movement in the 1970s gave birth to an alternative perspective towards development – the pluralist perspective. Mass communication was accorded a supportive and less directive role in assisting self-development activities of the local communities. The pluralist perspective advocated a growth of dualistic communication structure combining mass media at the national level with traditional media at the local level. Communication was seen as two-way, interactive and participatory.

Servaes (1991) argued that both the modernization and pluralist paradigms were still practiced in the developing world even though the former could not establish itself as a well-developed theory and the latter was attacked for its utopianism. With the advent of the information revolution, there was a resurgence of enthusiasm among the developing countries with hopes of using new ICTs to leapfrog the intermediate stages of technological development and participate in the global knowledge economy.

The literature suggests two contrasting views in terms of the relationship between ICTs and development in general and the Internet and rural development in particular. The optimistic view holds that the architecture and technological characteristics of the Internet promise rural people wider opportunities for alleviating poverty, enhancing education, promoting gender equality, improving health and encouraging participation. The international organizations and donor agencies strongly urge the developing nations to harness the benefits of ICTs to achieve their development goals. The dominant thinking is that the cost of investment in building ICT infrastructures in developing countries is high but the cost of not doing so could be much higher (Mansell & Wehn, 1998). The ICT-driven economic growth enjoyed by the U.S. and some OECD countries has generated much momentum for developing countries to invest heavily in national information infrastructures.

The pessimistic school pay more attention to the downside of ICTs such as the disparities resulting from unbalanced access and distribution of the benefits not only between the industrialized and the developing world but also among social segments within countries. They contended that the introduction of technologies into a society tended to consolidate the positions of the privileged and make the rich richer and the poor poorer. Brown (2001) and Heeks (1999) expressed their scepticism regarding the actual contribution of ICTs to poverty reduction. Kenny (2003) argued that the present understanding about the relationship between ICTs and social and economic growth was at a premature stage and there was a lack of evidence suggesting the cost-effectiveness of ICT investment in developing countries.

5.3 Research Questions

Several research gaps have been identified through the literature review, which help to formulate the research questions of the study. First, one of the conspicuous flaws with development communication research is its search for a theory of universal applicability, which contributes to either a total neglect or under-consideration of the structural constraints that developing countries are confronted with and an exaggerated role of communication and communication technologies in bringing about development.

The different approaches the three main paradigms have taken towards the conception of development, the role of communication in development and solutions to underdevelopment have both strengths and limitations. Given the fact that countries vary greatly from one another in terms of political, historical, and socioeconomic factors, treating developing countries as a homogeneous group and prescribing panacea for universal application is doomed to be fruitless. Therefore, the complicated nature of development goes against the pursuit of general patterns.

The predicament of development communication research was accurately captured by Okigbo's (1985) statement:

> Scholars have found themselves in a vicious cycle of self-defeating prophecy: poor results in efforts to formulate a theory of development based on the models of agricultural extension and knowledge gap lead to abandonment of further efforts to research communication and development; such an abandonment leads to even poorer results and much poorer research efforts, giving rise to a revolution of less efforts and less results, and consequently removing scholars farther from formulating a grand theory of development communication (p.24).

Jayaweera (1987) argued that the failure of the modernization paradigm was caused not by communication technologies themselves but the ill matching between the advanced technologies and the socioeconomic structures of the developing countries. In other words, communication technologies cannot substitute the appropriate policies for structural reforms. Goonasekera (1987) pointed out that mass media had to operate within a social system embedded with values, stratification, and political and economic conflicts. Hence, the nature of development is not necessarily a technical one; instead it can be a political one and depends on the willingness of the government to bring about social changes.

Moreover, communication has been accorded too powerful a role to be achieved by itself in the development process. Krippendorff (1979) argued that people in the developing countries were taken as a homogeneous group who were passive and inert, and readily to be influenced by the new media. Once communication was in place, it could set in motion the engines of development.

The same thinking contributes to the resurgence of the modernization paradigm in the last two decades of the 20th century and explains why there is so much hype about new ICTs in terms of their potentials in enabling developing countries to achieve the leapfrogging attempt. Jayaweera (1987) warned that developing countries started to replicate their mistakes in the 1950s and 1960s by expecting that satellite communication could bring about development that had not been achieved through television and radio. Consequently, development communication research will be coming around a full circle if it continues to be preoccupied with technologies per se and neglects the structural constraints of development. Therefore, more efforts should be devoted to understanding and analyzing the complex nature of societies and cultures in which development is supposed to take place rather than seeking a "preordained direction" of development for generalization (Schramm, 1976, p.47).

Second, the positivist approach adopted by development communication researchers and their preoccupation with individual changes lead to the quest for causal relationships. The hypothesis central to development communication is that communication technologies are important causative factors inducing and resulting in positive social changes. This research orientation has guided many of the classical studies towards seeking and predicting the causality between communication and development variables.

For example, the modernization theory was considered as a mere historical observation and description of the route of Western modernization. Lerner's (1958) proposition about the components and sequence of modernization was assumed rather than empirically verified. His findings based on countries' aggregate data only suggested strong correlations between every set of two variables (media participation and literacy; literacy and urbanization; literacy and voting, industrialization and literacy, literacy and per capita income). However, correlations could not help to indicate the sequence of occurrence.

Rogers and Svenning's (1969) meta-analysis of several studies of modernization among peasants also failed to prove the antecedent and consequent paradigms they conceptualized to outline the modernization process. As the co-varying relationships between variables did not reveal their time order or causality, Rogers and Svenning concluded that modernization was a multivariate process where a multitude of variables were involved in a web of interdependent relationships.

A recent attempt to establish a causal relationship between ICTs and social development by using data from 175 countries in the *Human development Report 2003* did not succeed in substantiating a significant direct linkage either (Hao & Zhao, 2004b). The study concluded that the utilization rates of ICTs might serve as indicators of a country's social development but there was weak evidence to demonstrate the role of ICTs in directly promoting social development.

The search for generality and causality reflects the tendency of simplifying the complicated development process among development communication scholars. Jayaweera (1987) argued that "because development itself is a function of a whole range of other variables, the precise correlations between DC (development communication) and development on the one hand and between DC and the other development variables on the other, have been difficult to research or evaluate" (p.77).

The intertwined and indefinite relationship between communication and development in general and ICTs and rural development in particular reorients us towards alternative perspectives other than the pursuit of a precise causal relationship. A more in-depth understanding about how communication is involved and used in the process of development instead of the attempt of pinpointing which part of development is exclusively caused by communication and which is not, will better enhance our existing knowledge about development communication.

Third, the pro-innovation bias of diffusion research has resulted in the extensive testing of individual innovativeness, while the influences of social structures on the diffusion process, the social consequences brought by innovation adoption, and the possible discontinued use of innovations remain understudied.

The diffusion of innovations theory was formulated and developed on the basis of observing and analyzing the practices of American agricultural extension under a well-developed market system. Though innovative ideas or change forces are introduced, the decision of adoption or rejection is more often determined by the will of individuals. Most diffusion studies focus on the social and individual traits of adopters and the roles that mass media and interpersonal communication play at different stages of decision-making process. The dominant research trend overlooks the collective or organizational efforts in extending innovations among the targeted audiences, especially when conditions are unfavourable to individual adoption of innovations.

Under the influence of pro-innovation bias, the diffusion studies often stop its discussion at the adoption stage and most extension activities aim at speeding up the adoption rate. As innovations tend to bring about both positive and negative consequences and adopters may stop using an innovation when driven by less perceived needs, the new research perspective calls for identification of factors that influence actual usage of innovations at the post-adoption stage and appreciation of the implications of innovations from users' reference points.

Fourth, the extant literature suggests a lack of reliable data regarding the linkages between the Internet and rural development. The romanticization of the potential of ICTs in solving underdevelopment problems has caused the mixing up between what the Internet can be expected to achieve and what has actually been achieved. Moreover, most of the empirical studies are anecdotal in nature and lack theoretical underpinnings and sound methodological design.

The identification of the research gaps helps to provide directions for the present study. The Internet diffusion and usage and the possible development implications should be examined and discussed by taking the political, socioeconomic, and technological contexts of rural China as frames of reference. Instead of seeking a development model for generalization, the study aims to identify specific features of the diffusion process that are unique to the Internet (in contrast to other technologies), the Chinese context (in contrast to other countries), to the rural users (in contrast to urbanites), to rural areas with various levels of development and geographic conditions.

In view of the inconclusive findings on the role of communication in bringing about development, the study takes a qualitative approach by viewing the diffusion and usage of the Internet in rural areas as a whole process. It seeks an in-depth understanding about how the various contextual factors condition the diffusion and adoption of the Internet among rural people and the possible implications of the technology on their life.

The shortage of reliable data to validate the potential of the Internet in enhancing rural development suggests that more efforts should be spent in understanding the impacts of the Internet on the livelihood of the rural people, the various parties involved in Internet diffusing initiatives, the cost-effectiveness or financial viability of Internet services, the socioeconomic benefits derived from improved access to information via the Internet, the relevance of Internet content to the needs of farmers, and the complimentary role of the Internet to traditional means of communication.

In view of the above discussion, this study sets out to answer the following research questions:

RQ1: How are Internet diffusion and usage shaped by the political, socioeconomic, and technological contexts of rural China?

RQ2: What implications do Internet diffusion and usage have upon various aspects of rural development?

RQ3: How do governments, the private sector, as well as individuals contribute to Internet diffusion and usage? What are the possible ways of providing a viable Internet service to rural areas?

5.4 Summary

The literature review shows gaps in development communication research in general and diffusion of innovations studies in particular. The search for theories of universal applicability, the neglect of structural constraints, the dominance of positivist approach, and the assumed rather than empirically verified causal relationship between communication and development are identified as the main reasons resulting in the lack of progress in development communication.

To fill the research gap, this study adopts a qualitative approach by taking Internet diffusion and adoption as a social process. Instead of seeking for universal applicability, the study explores the diffusion and usage patterns of the Internet as well as its implications on development under the specific context of rural China.

The following chapter provides a detailed description of the methodology that the study employs to answer the research questions, including specific methods used in data collection, selection of research settings, sampling procedures, and data analysis techniques.

6. Methodology

6.1 The Qualitative Research Paradigm

In view of the exploratory nature of the present study, I adopted a qualitative research paradigm to seek answers to the research questions. Research paradigms reflect assumptions that researchers take towards the nature of social reality and prescribe the selection of certain research methods, which in turn lead to different understanding and interpretation about social phenomena.

While the quantitative approach perceives the social world as an objective reality and is concerned with making predictions through generating universal laws between the component parts, the qualitative perspective believes that society is constructed through the subjective experiences of its members and intends to "produce a unique explanation about a given situation or individual" (Wimmer & Dominick, 2000, p.104).

The most typical qualitative assumptions are summarized as follows, which distinguish themselves from those of the quantitative methodology.

1. Qualitative research is conducted in everyday natural settings, concentrating on the natural occurrence of human behaviours and events (Creswell, 1994; Wimmer & Dominick, 2000).
2. The researcher becomes the primary instrument for data collection and analysis and the active involvement of the researcher in data gathering process makes him an integral part of the data. In a qualitative study, the researcher more often than not gets himself immersed into the daily life of the social group under study and interacts with them to observe and

record their sociocultural practices (Eisner, 1991; Lincoln & Guba, 1985; Merriam, 1988).

3. Qualitative research focuses on the process rather than the outcome. Researchers attempt to find out how and why things occur (Merriam, 1988).
4. Qualitative research is exploratory in nature, and more amenable to investigating social phenomena when variables and theory base have not yet been clearly identified (Merriam, 1988).
5. Qualitative research centres on how the participants perceive and make sense of their lives. Through reconstructing the social realities in the eyes of the participants under study, the researcher seeks to create meaning, generate insights, and explain events (Lincoln & Guba, 1985).
6. Qualitative research uses an inductive instead of a deductive approach. Theories evolve from the data collected and are not established prior to studies (Merriam, 1988; Wimmer & Dominick, 2000).
7. Qualitative research is guided by purposeful sampling – "sites or cases are chosen because there may be good reason to believe that 'what goes on there' is critical to understanding some process or concept, or to testing or elaborating some established theory" (Schwandt, 1997, p.128).

The rationale for selecting the qualitative research paradigm to dictate the research design of the present study has been established in Chapter 5. I find Morse' (1991) exposition quite helpful in explaining and justifying the selection. Morse stated:

> Characteristics of a qualitative research problem are: (a) the concept is "immature" due to a conspicuous lack of theory and previous research; (b) a notion that the available theory may be inaccurate, inappropriate, incorrect, or biased; (c) a need exists to explore and describe the phenomena and to develop theory; or (d) the nature of the phenomenon may not be suited to quantitative measures (p.120).

As aforementioned in the literature review, three decades of development communication research has gone into a stagnant status due to the failure of discovering a grand theory of universal applicability. Fewer efforts have been spent on theoretical formulations about the impact of communication on development thereafter because it is an "over-ploughed, yet under-yielding field" (Okigbo, 1985, p.24). The absence of theories in guiding future research leads to the phenomenon that most of the roles of communication and communication technologies in enhancing development are assumed rather than empirically verified. This is evidenced in the striking gap between the much hype around the potentials of ICTs in helping developing countries achieve their development goals and what changes they have actually brought about in the developing world.

Schramm (1976) realized that the complex nature of societies and cultures did not ensure some "preordained direction" of development (p.47) and it was not possible to seek generality regarding the relationship between communication and development. This generates the need to explore the implications of Internet diffusion and usage on development within the specific context of rural China.

The positivist origin in development communication research in general and diffusion of innovations studies in particular has contributed to the dominance of theory-testing model aiming to verify the hypothesized causal relationship between development and communication variables. The majority of diffusion studies centre on the individual innovativeness prediction and decision-making process, with less consideration for the social system where the act of adoption is supposed to occur.

However, as development is considered a multivariate process in which a multitude of concepts are involved in a web of interdependent relationships (Rogers & Svenning, 1969), this renders a quantitative approach unsuitable for the study on the linkage between communication and development. This has been substantiated by the pitfalls of the modernization research in establishing causal relationships and the limited progress made by diffusion studies since the findings only stay at the level of description and categorization.

Edelstein (1984) pointed out that the research methodology employed by the modernization school should be improved because "generalizations should emerge from data rather than data being used to test preconceived ideas that were neither scientific nor empirical in origin" (p.48). This is proposing that the method of induction should be more appropriate for development communication research. Therefore, this study took a qualitative approach to seek alternative understanding about the possible linkages between the Internet and rural development.

6.2 Data Collection and Data Analysis

As the main purpose of the present study is to explore ways that the Internet is employed to promote social and economic development in rural China, it adopted the ethnographic research design, which enabled the researcher to interact with the rural people under study and to observe and record the daily occurrence of events.[8] In-depth interviews, participant observations and documentary analysis were used to seek answers to the research questions. However, in view of the time constraints and the limited degree of immersion into the daily life of the participants, this research attempt could not qualify itself as an ethnographic study in a strict sense. Therefore, I would like to claim that ethnographic techniques were utilized in the research design to guide data collection and analytical procedures.

8 Ethnography is a research methodology associated with anthropology and sociology that systematically describes the culture of a group of people. The goal of ethnographic research is to understand the natives' or insiders' view of their own world. Spradley (1979) held that "People everywhere learn their culture by observing other people, listening to them, and then making inferences. The ethnographer employs this same process of going beyond what is seen and heard to infer what people know. It involves reasoning from evidence (what we perceive) or from premises (what we assume)" (p.8).

6.2.1 *The Research Settings*

In order to obtain a relatively comprehensive understanding of the phenomenon under investigation, the strategy of maximum variation sampling was used in selecting research settings. The study was conducted in five rural places, the selection of which was based on (a) the dominant parties in initiating the Internet diffusion and adoption among rural people and (b) the specific contexts conditioned by various levels of socioeconomic development and geographical locations. The classifications were meant to represent the prevalent Internet diffusion models in China and therefore generate meaningful comparisons.

a) Yellow Sheep River Town, Gansu Province

Yellow Sheep River is an impoverished mountainous town in Gulang County, Gansu Province. The Internet was introduced to the Yellow Sheep River vocational school by a Taiwanese company in 2000 with the hope of popularizing computers and network technology and fostering knowledge-based economic development there. The experiences of Yellow Sheep River stand for the private initiative to use the Internet in support of rural development (see Table 2 and Figure 4).

b) Jinta County, Gansu Province

Jinta is located in the northeastern part of Jiuquan City, Gansu Province. The county government has carried out an ICT project called "Networking the Villages and Informing the Farming Households" since 2001 to build the last mile connectivity through providing farmers with information over the Internet. The experimental project at Jinta County has demonstrated how the Internet can be embraced by farmers through planned efforts of the government (see Table 2 and Figure 4).

c) Tongnan County, Chongqing Municipality[9]

Tongnan County lies in the northwest of Chongqing Municipality. It is among five poor counties covered by the Project on Poverty-Reduction through Access to ICTs, jointly sponsored by the Chinese Ministry of Science and Technology (MOST) and UNDP. This is a unique case of Internet diffusion initiative as a result of cooperation between central and local governments and an international organization (see Table 2 and Figure 4).

d) Pinggu District, Beijing Municipality

Pinggu is a suburban district in the northeast of Beijing. The local government has initiated an information network project to connect rural people with the Internet since 2000. The Internet has been used as an efficient tool to spread E-literacy among local farmers over the years. As a rural area with middle level of socioeconomic development, Pinggu is another example showing the role of government in promoting Internet diffusion and usage (see Table 2 and Figure 4).

e) Shouguang City and Huanglou Town, Shandong Province

Shouguang City and Huanglou Town are situated in Shandong Province, with the former known for its greenhouse vegetables and the latter for its flower production. The local farmers are better off as compared with those in western or inland areas. This gives them opportunities of accessing the Internet on their own initiatives since they can afford the new technology. Three rural entrepreneurs' experiences of adopting and using the Internet provide a better understanding about farmers' Internet usage and its implications on rural development

9 The PRC administers 33 province-level divisions, including 22 provinces, five autonomous regions, four municipalities, and two special administrative regions. Beijing and Chongqing are among the four municipalities. There are five practical levels of local government in mainland China: the province, prefecture, county, township, and village.

when the Internet diffusion and adoption occurs in a relatively natural state (see Table 2 and Figure 4).

Table *2*
Five Rural Places Selected for the Study

Diffusion Models	**Selected Sites**	**Location**
Private Organization	Yellow Sheep River	Western
Government	Jinta	Western
Government + International Organization	Tongnan	Western
Government	Pinggu	Central
Rural Entrepreneur	Shouguang & Huanglou	Eastern

6.2.2 Data Collection

Data collection took place from April to August 2005. The actual stay in each of the five rural sites was around 20 days, with the rest of the time spent in negotiating access and travelling. The main data collection methods included in-depth interviews, participant observations and documentary analysis.

a) Negotiating Access

The permission for carrying out studies at the five research sites was obtained through contacting local governments at the county or city level. As field research is rarely conducted in rural areas, winning government support can facilitate the process of data collection and grant legitimacy for the study. To a certain extent, it is hard to talk rural people into believing that a researcher travels over long distances just to observe and understand their life. Therefore, I tried my best not to arouse suspicion and misunderstanding so that a rapport could be developed with the local residents.

Figure 4. Five rural places selected for the study.[10]

b) In-depth Interviews

Face-to-face interviews were arranged with three groups of informants – officials at various government levels, information technicians maintaining Internet services and farmers with access to the Internet (or online information). The strategies of maximum variation and snowball sampling were employed in the selection of the 67 interviewees.

At each research site, government officials in charge of planning and implementing ICT projects were interviewed. The technicians in

10 A. Jinta B. Yellow Sheep River C. Tongnan D. Pinggu E. Shouguang & Huanglou

charge of Internet services were also approached for interviews. In turn, these informants recommended interviewees among local Internet users because they had a good knowledge about the regular patrons of information centres and those who had successful experiences of using the Internet (or online information) for life enhancement.

In the sampling of interviewees, efforts were also made to strike a relative balance between rural residents whose occupations did not involve farming (such as village committee members, school teachers and students) and farmers directly engaged in agricultural production or marketing. The main sampling criterion had more to do with information-richness of the cases than with the sample size.

Most of the interviews were conducted in the working environment of the informants or at places where they could access the Internet because this enabled me to better understand the social context. In her study about the Internet and construction of Chinese cultural identity, Chan (2004) argued that the artefacts in the living and working environment of the interviewee might give more information about his life attitudes.

All the interviews were conducted in Chinese and recorded after getting informed consent of the interviewees. Verbatim transcriptions were made for the convenience of analysis. These interviews totalled about 45 hours of taped conversations and 150 pages of transcriptions.

c) Participant Observations

As there is a wide range of roles for researchers who engage in participant observation, it is important to point out that my role in the present study is more approximate to an observer-as-participant. Gold (1958) developed a typology comprising four kinds of observer roles: the complete participant, the participant-as-observer, the observer-as-participant and the complete observer. The complete participant is recognized as a member of the group or organization under study and the researcher's identity is not known. Those under study are not aware that they are being observed. The participant-as-observer is often assigned a part-time or temporary role by the gate-keeper, who gives the researcher permission to conduct research in the setting. The observer-as-participant is more concerned about his or her role as an observer and refrains from participating in the process under observation. The complete observer can be considered as a covert observer, who monitors the activity of a group from the sideline and does not engage in interaction with those under study.

My background as an urban dweller for over 30 years makes it hard for me to get immersed in lives of the villagers and really become one of them. This is another reason why the research could not qualify itself as an ethnographic study. The limited interaction with the participants made me fall short of the requirements of an ethnographer. Therefore, I was more prone to the role of an observer and less to the role as a participant.

The focus of observation was everyday experiences of farmer uers and daily events at information centres. Particular attention was paid to Internet access mode and use patterns, frequently used functions and websites, and the ways the Internet, especially through the information conveyed and the communication enabled by the technology, could possibly influence the lives of rural people. In addition, the daily management of information centres as well as the technical support offered were also attended to.

Field notes were taken and a field dairy was kept to chronicle the daily experiences, feelings and perceptions. Digital cameras proved to be an efficient tool to quickly capture texts (such as wall posters) and events especially when time was limited for observation. Pictures of interviewees and their living environments were also taken with prior consent. As precious documents of life, these digital pictures helped to relive my experiences even after exiting the research sites.

d) Document Collection

The existing documents provide extra information about a setting or groups of people, which do not lend themselves to direct observation and questioning (Lincoln & Guba, 1985). In the study, public records such as county level censuses, statistics reports, county office records and government circulars were read for contextual understanding about levels of socioeconomic development and rural telecommunication infrastructures. Such materials were also helpful in making comparisons among research sites.

With regard to ICT projects carried out at the selected rural places, internal records such as project implementation reports, minutes of meetings, presentations, logs about Internet access and ICT training materials were collected to supplement the data gathered through interviews and observations. In addition, books, newspaper clips and audio visual materials featuring interviewees were used as personal documents to help me better understand their lives and appreciate the meaning of the Internet from farmers' point of view. After getting institutional or personal consent, paper documents were photocopied and electronic files were stored on flash drives for easy carrying.

6.2.3 Data Analysis and Interpretation

Despite the electiveness in terms of analysis of qualitative data, there are some common procedures taken by qualitative researchers (Lindlof & Taylor, 2002; Tesch, 1990). The qualitative data analysis involves data "reduction" and "interpretation" (Marshall & Rossman, 1989, p.114). If the data reduction process can be considered as "decontextualization", then the interpretation means "re-contextualization" because "while much work in the analysis process consists of 'taking apart' (for instance, into smaller pieces), the final goal is the emergence of a larger, consolidated picture" (Tesch, 1990, p.97).

For this study, the data analysis started from the transcription stage. I wrote down the reflective notes about what I learned from the data while listening to the recordings and recalling the scenes where interviewees made the comments. Thus my first reaction to the discourses as well as the fleeting ideas was recorded to offer clues for further conceptualization. The interviews, field notes, field diaries and documents were typed into word processing documents and transcribed data of each research setting were treated as a separate data set so that comparison could be made through figuring out similarities and differences.

Next was the coding stage, which meant whenever I came across a meaningful segment of text in a transcript, I assigned a category name to signify that particular segment. This procedure allowed me to reduce the overwhelming amount of data and sharpen the focus of analysis onto meaningful analytical units with reference to the research questions. For example, in the data set of Yellow Sheep River, there were several incidents showing how the Internet was used by local farmers. The owner of the clinic mentioned that the E-centre helped him sell medical herbs collected from the mountain. I categorized this as "Locating potential buyers". Another event showed that with the help of the E-centre, a local teacher received donation and encouragement over the Internet. This was coded as "Winning social support".

The "constant comparative technique" was employed to identify and describe some recurrent topics from the perspectives of the participants (Wimmer & Dominick, 2000, p.107). After repeated reviews of data, the texts which clustered around the same topic were sorted out and labelled under the same category. This also required the constant refinement and elaboration of categories until an exact definition of each category and rules governing the inclusion of incidents were established to narrow the focus of the study.

Following the initial coding, a master list was kept with all the codes developed and used in the study. The next step involved identifying relationships among categories and organizing them into different levels. For instance, I found "Locating potential buyers", "Setting a market value for land produces", and "Making informed decision on buying and selling" revealed the concrete ways in which the Internet helped with trading of farm produces. Therefore, a theme named "Facilitating transactions of farm produces" was developed. At this stage, interrelationships and common regularities and patterns were derived across categories.

At the final stage, I integrated the themes into a conceptual schema that tied my observations about the transcribed data together to offer a logical interpretation of the phenomenon under study. The experiences and perceptions of the informants were reconstructed through the report of summarized findings.

As a supplement to the qualitative data analysis, the method of content analysis was used to examine a local newspaper based on online information in Jinta County. Every online issue available was downloaded and analyzed to provide an objective and quantitative description of various kinds of agricultural information gleaned from the Internet. The analysis offered a clear understanding about the close linkages between newspaper content and local farming cycles. This helped to enrich the discussion about the importance of content relevance with regard to the diffusion and use of online information among local farmers.

6.3 Ethical Considerations

In this section, I am going to discuss the ethical issues in association with the present study and the safeguarding measures taken to resolve them. The ethnographic methods can be quite intrusive to the life of the people under study. As most of the rural communities are conservative and ill-informed, the presence of an outsider could bring a lot of suspense and uneasiness among the participants. They might feel that something goes wrong and the researcher is someone sent to gather evidence.

Another ethical concern arises from the sensitive information the informants might reveal to the researcher regarding the assessment of the actual contribution of the Internet towards their life. In some pilot ICT villages, the negative feedbacks about government-supported information centres might have an adverse effect on the informants if they were made public. Hence, there was a need to protect the privacy of the informants.

In order to address the possible ethical concerns, informed consent was obtained from government officials and farmers under study. The informants were provided detailed information on research objectives as well as data collection devices and activities (both in written form and verbally) before they made decisions on whether to take part in the study.

With regard to the protection of privacy, the verbatim transcription and the final research report were made available to the informants. The rights and desires of the informants were taken as my top concern when reporting the data. The informants had the final say on whether he should be kept anonymous or not.

6.4 Reliability and Validity

Though the concepts of reliability and validity bear different connotations for qualitative research in comparison with the quantitative one, they are nonetheless essential issues that directly determine how much faith we should place in a researcher's conclusions. Qualitative researchers have developed their own interpretation and criteria to build credibility of their studies.

According to Merriam (1988) and Miles and Huberman (1984), the issue of internal validity in qualitative research emphasizes the trustworthiness of the information and its dovetailing with the reality. The following strategies were taken to ensure internal validity of this study:

1. Triangulation of data. The use of in-depth interviews along with participant observations and documentary analysis enabled the problem under study to be examined from different perspectives. As Emmert (1970) put it, "Since each measurement method has its own bias and distortions, a variety of

instruments may mean a more complete picture of the object of investigation" (p.163). Therefore, each method served as a complement to others. I came to know what was happening through posing questions to informants in in-depth interviews and participant observations helped me understand and explain what the informants said. Moreover, the interviews conducted among informants of various social roles in this study also contributed to the verification of the collected data.

2. Audit Trail. All the original data for analysis and the analytical methods used were kept for others to examine the thinking process involved in the work.
3. Member checks. The interview transcripts, field notes and categories and themes identified were given to the informants and participants to make sure that I had accurately interpreted what they said and what the phenomenon was like.

Though the intent of the present study was not to generalize the findings to other parts of rural China or other developing countries and its uniqueness within the specific context of rural China did not allow it to be exactly replicable in another context, limited external validity and reliability could be guaranteed through providing a detailed account of sampling procedures, data collection methods, data analysis process as well as reflexivity of the researcher (in next section). These endeavours enhanced the chances of this study to be replicated in another setting. The multiple methods of data collection and analysis also strengthened the reliability of the study.

6.5 Reflexivity

As the researcher acts as the primary data collection and analysis instrument in the qualitative research, it is necessary to identify his educational background, personal bias and philosophical views, which may shape the understanding and interpretation about the setting and people under study. Reflexivity means the researcher's self-consciousness about his presuppositions, biases or personal values that may keep him from making objective and value-free judgments (Brewer, 2000; Hammersley & Atkinson, 1995).

Most of my preconceptions about rural people and their life originate from books and films. As I have been brought up in an urban environment and there are few opportunities for me to know about what the countryside looks like, my limited understanding is only built upon several excursions and vicarious experiences from the mass media. In view of the sharp economic and social disparities between urban and rural China, it is not exaggerating to say that I, as an educated urban dweller, and my informants and participants, are living in two completely different worlds. It is a great challenge for me to build rapport with rural people and appreciate their situation and needs.

However, a preliminary study on cyber communities I conducted at Pinggu District (the suburban area in Beijing) in 2003 allowed me to accumulate some experiences in communicating with rural people. In the course of the study, questionnaire surveys were carried out in three villages and local officials and village committee members were interviewed. This helped me foresee some of the problems that might crop up in data collection and seek satisfactory solutions.

6.6 Summary

The research questions posed by the study prescribe that a qualitative research paradigm should be taken to seek better understanding and interpretation about the possible impact of the Internet on rural development. This study adopted an ethnographic research design, with in-depth interviews, participant observations, and documentary analysis as the main data collection methods. Five research sites of varying geographic locations and social and economic development levels were selected based on the prevalent Internet diffusion models in rural China.

This chapter has also given a detailed account of the procedures and strategies taken in data collection and data analysis, such as negotiating access, sampling procedures for interviews, and data analytical tools. The ethical problems associated with the study have been addressed and concrete measures have been taken to protect the privacy of the informants and minimize the possible harm incurred by the intrusive data collection process.

In terms of the reliability and validity of the study, the strategies of triangulation of data, audit trail and member checks have been adopted to ensure the internal validity. Limited external validity and reliability can be guaranteed through a minute account of the sampling procedures, data collection methods and data analysis processes. In the section of reflexivity, the researcher mentioned that the social distance between herself and rural people became a major impediment in arriving at objective and value-free judgments about the population under study.

7. Rural China as a Testing Ground

China has experienced one of the fastest rates of economic growth in the 25 years between 1978 and 2003, with the annual growth rate standing at an average of 9.4% and its GDP rising from 147.3 billion US dollars to over 1.4 trillion US dollars (Hu, 2004). Generally speaking, this signifies the enhancement of living standards for a large portion of humanity since the country has almost one-fifth of the world population.

Benefiting from the robust economic development, China has made remarkable progress in poverty reduction. The number of rural people living in absolute poverty decreased from 250 million in 1978 to 26.1 million in 2004 (IOSC, 2001; NBSC, 2005).[11] If measured by the stringent World Bank standard of one US dollar daily income, the impoverishment rate in rural areas declined from 33% in 1990 to 16% in 2000 (UNDP, 2003). Much of the impressive reduction in global poverty in the 1990s is due to the systematic efforts China made in the sector (World Bank, 2001).

China's per capita GDP reached a record high of 1,490 US dollars in 2004. Economists predict that China has entered into a turning point in its development, with the per capita GDP reaching between 1,000 and 3,000 US dollars (Cui, Lu & Wang, 2005). Like many other transitional societies, a number of social problems have cropped up in the course of rapid economic development, including rising unemploy-

11 The Chinese government takes the lowest expense to maintain one's basic subsistence as the standard of poverty. The poverty line in 1985 was RMB 206 (U.S.$25) in per capita income in rural areas. The readjusted standard in 2004 was RMB 668 (U.S.$84) in keeping with the economic development (IOSC, 2001; NBSC, 2005). The official poverty line is much lower than the standard of one US dollar daily income set by the World Bank, equal to RMB1, 000 measured on a purchasing power parity (PPP) basis.

ment, widening gap between the rich and the poor, and great social instability. In particular, the issues concerning rural areas, agriculture and farmers are among the top concerns of the Chinese government. This chapter aims to provide sufficient background information about the socioeconomic, political and technological contexts of rural China and thereby set up a frame of reference with respect to the diffusion of the Internet and its implications on rural development.

7.1 An Overview of Rural Issues

According to the National Bureau of Statistics of China, there are 757 million people (around 58.2% of the total population) living in rural areas by the end of 2004 (NBSC, 2005). In terms of political divisions, the country is divided into five administrative levels: provinces, cities, counties, townships, and villages. At the provincial level, the central government administers 22 provinces, five autonomous regions, four municipalities, and two special administrative regions (Hong Kong and Macau). Whereas townships are the lowest level of administration with political power, villages act as the basic autonomous organizations among rural people.

7.1.1 Rural Reforms

The rural reforms in the late 1970s and 1980s have had a significant impact on the social and economic structures of the contemporary rural society. The Third Party Plenary Session of the 11th Party Central Committee began to reform the rural economic structure by replacing the commune system with the household contract responsibility system. Previously, the communes or the subordinate production teams were the basic units of rural administration and economic operation. The land was collectively owned by the production teams, which made decisions about planning production, sharing land produces, and distributing incomes.

By contracting the land to each rural household, farmers were given the right to make their own production decisions and take full responsibility for profits or losses, which used to be shared among members of the collectives under the commune system. Farmers were allowed to consume the surplus of land output or sell it in rural markets after they delivered the crop quotas set by the state (Delman, 1993). The new land system boosted farmers' production incentives and encouraged them to demonstrate their capabilities, and use resources and skills for their own benefits.

The overall reform of the commune system began in 1982 and around 95% of rural households were under the household contract responsibility system by the end of 1984. Along with the introduction of the new land system, those policies restricting agricultural development were abolished. The state monopoly of agricultural products and rural market was ended and farmers were allowed to engage in diversified business and set up their own enterprises (Lu, 2002).

The new rural policies contributed to a steady growth of agricultural productivity and a more diversified rural economy. The gross value of agriculture output in 1985 was almost twice as much as that of 1978 and the production of grain, oilseeds, cotton and livestock rose rapidly. The per capita net income of rural households nearly tripled in this period, increasing from RMB134 in 1978 to RMB 397 in 1985 (FRDLC, 2005). With the reopening of rural markets and the resumption of business activities, the village and town enterprises started to flourish. These grassroots industry assimilated much surplus labour from the countryside. The rural economic structure underwent a rapid transformation, with the ratio of output value between agriculture and non-farming industries changing from 7:3 in the late 1970s to 3:7 in the mid-1980s (Fei, 1989).

Despite its contribution to the revitalization of the rural economy, the household contract responsibility system has its inherent drawbacks. With the basic economic units shifting from production teams to individual rural households, the village collectives stop functioning as production planners and organizers and therefore have less control and influence over the rural inhabitants (Ye, 2002). Driven by a strong desire to get rich, farmers tend to focus on the productivity of their contracted lands and spend less effort in building public facilities. The village organizations find it difficult to organize and mobilize farmers to construct and maintain public infrastructures, such as schools, roads and irrigation systems (Liu, Meinzen-Dick, Qian, Zhang, & Jiang, 2005). For rural areas with low collective economic capacity, the public services and welfares have not undergone much improvement, despite the overall increase in household incomes.

As a direct result of the responsibility system, over two million individual household production units in agriculture, sideline production and market activity have emerged. The small-scale household economy often operates with very little capital and remains disorganized. Due to a lack of collective institutions to provide pre- and post-production services, each farming household has to represent itself at the market (Lu, 2002). The ill informed farming decisions and backward production modes often put farmers in a disadvantaged position in a market economy based on supply and demand.

The high degree of farm fragmentation also poses difficulty for the diffusion of new agricultural technologies and the implementation of mechanical farming. Z. Wang (2005) documented that the average contracted land area was only 0.38 hectares per household in 2000. The small sized farms prevent rural households from using modern machineries and make them less likely to realize production gains from scale economies (Carter & Estrin, 2001).

7.1.2 The Urban and Rural Divide and Socioeconomic Segmentation of Rural People

China has been under a rigid dual structure of urban and rural division since the adoption of the household registration system (*Hukou*) in the mid 1950s. The system divides all Chinese people into agricultural population and non-agricultural population and serves as an efficient control mechanism of rural-urban migration. Under the system, non-agricultural population or urban residents enjoy many social benefits such as housing, employment, education and medical services, while farmers are denied the benefits available to the urbanites even if they move to live and work in urban areas (Li, 2004). Therefore, the household registration system is in effect a very important form of social closure[12] that excludes the overwhelming majority of Chinese people from sharing urban resources.

The pre-existing socioeconomic disparities between urban and rural areas have not been alleviated along with the rapid GDP growth. China has witnessed a widening gap between urban and rural incomes in recent years even though the livelihoods of rural people improve in general. With the per capita annual income of rural people reaching RMB 2,936 (U.S.$367) in 2004, China has become one of the countries with the largest urban-rural divide in the world. The average per capita income of urban residents was 3.2 times that of rural dwellers, much higher than the corresponding figure (1.7 times) in 1984 (Lu, 2002; NBSC, 2005). In fact, the real gap can be even wider considering the social welfares made available for the urban people. The sharp urban-rural inequality has posed direct threats to China's social stability and greatly impeded its sustainable development.

With the ongoing economic reforms entering its third decade, China has also had a sharp increase in income inequality among rural people not only at different regions but also within the same region. Farmers' per capita net incomes at the eastern, central and western regions [13] stood at RMB 3,616.6 (U.S.$452), RMB 2,382.1 (U.S.$ 298) and RMB1,878.9 (U.S.$ 235) respectively in 2003 (MOST, 2005a).

12 Social closure, a notion raised by Max Weber, is used here to refer to the degree to which social classes are isolated from one another (Murphy, 1990).

13 The eastern region includes Beijing, Tianjin, Hebei, Liaoning, Shanghai, Jiangsu, Zhejiang, Fujian, Shandong, Guangdong, Guangxi, and Hainan. The central region includes Shanxi, Inner Mongolia, Jilin, Heilongjiang, Anhui, Jiangxi, Henan, Hubei, and Hunan. The western region includes Chongqing, Sichuan, Guizhou, Yunnan, Tibet, Shanxi, Gansu, Qinghai, Ningxia, and Xinjiang (MOST, 2005a).

Generally speaking, farmers living at the coastal provinces have more disposable income than their counterparts at the remote inland areas. However, the regional averages can hardly capture the socioeconomic differentiations within regions since China is a country so large in population and territory. There may be poor families living below the minimum subsistence level in fairly well-off areas, and vice versa.

The replacement of collective economy with household economy has allowed farmers to directly enjoy the fruit of their own labour, as the even distribution of land produces and collective revenues is replaced by a system in which the input and output in farming are linked. Furthermore, the government policy to encourage some people and some places to become rich first has also motivated farmers to strive for wealth (Delman, 1993).

Accordingly, the rural population has gradually moved "from being socio-economically homogeneous (at least within regions) to becoming segmented" (Delman, 1993, p.40). This has given rise to a new stratification among rural people. On the basis of their occupations and different interests and needs that they represent, rural people mainly fall into eight large groups: farmers, farmer workers, employees, rural intellectuals, individual industrialists and businessmen, private entrepreneurs, managers of township enterprises, and rural administrators (Lu, 2002; Zhu, 2002).[14] The socioeconomic segmenta-

14 According to Lu (2002), farmers account for 55%–57% of the rural people. Farmer workers, employees, rural intellectuals, individual industrialists and businessmen, private entrepreneurs, managers of township enterprises, and rural administrators account for 24%, 4%, 1.5%–2%, 5%, 0.1–0.2%, 3%, and 6% respectively.

Farmers refer to those rural people who take crop cultivation and livestock rearing as a chief means of living. Even though farmer workers may work in factories, they are still considered as agricultural population on the basis of the household registration system. Employees stand for those farmers who are employed by private businesses. Rural intellectuals remain relatively better educated as compared to the overwhelming majority of the rural people and they are often engaged in education, technology and cultural sectors. Individual industrialists and businessmen used to be capable craftsmen in the rural communities, such as carpenters, tailors and tilers. With the implementation of the household

tion of rural society has created conditions for varying groups to adopt the Internet for different development purposes.

7.2 China's Experiences of Mobilizing Communication for Rural Development

The Chinese government has accumulated rich experiences in using various means of communication to mobilize public support and direct social and economic changes. Ever since the foundation of the country, the Communist Party of China (CPC) has developed an efficient communication network integrating political institutions with media operations to exert control over mass media. The government has taken mass communication as important social resources to realize its development goals at various stages. Therefore, the diffusion of media technologies as well as the deployment of various means of communication is often a result of planned and purposive government initiatives. The historical shifts in rural communication have provided full support for this.

---

contract responsibility system, these people set up small-scaled family businesses. Private entrepreneurs refer to those rural people whose companies hire over eight employees. Managers of township enterprises organize and coordinate the production and selling in collectively owned enterprises in rural areas. Rural administrators include carders at the town and village levels, who are responsible for the social stability and economic development in rural areas.

7.2.1 Rural Communication before 1978

At the early stage of China's development, the building of a far-reaching broadcasting system was considered both a political and an economic necessity. When the new nation was founded, almost 80% to 85% of the population were illiterate. The government had to develop an extensive broadcasting system and take advantage of every potential of oral communication to bypass the widespread illiteracy (Liu, 1964).

The sound broadcasting service started in 1940 and became fully fledged in 1976. The system was composed of three parts: wireless broadcasting, wired broadcasting and a monitoring system. Wireless broadcasting referred to the central and provincial radio stations that used radio waves to transmit information to different relay stations, which in turn passed on the radio signals to individual radio sets. Wired broadcasting was used at the local level. It was "a point-to-point radio communication system which disseminated selected programmes at the point of reception through loudspeakers" (Lin, 1993, p.15). Radio broadcasting monitoring system served as an aid to ensure that important programmes (such as lectures by government leaders and special lessons on social science) were received by the population. The appointed monitors would introduce and announce the programmes broadcast by the central and provincial broadcast stations and organize local people to listen to these programmes (Liu, 1964). The integration of the three elements contributed to an extensive broadcasting network that could reach the majority of the public in the country (Lin, 1993).

The easy affordability and low maintenance costs of the wired radio broadcasting system made it ideal for rural areas where people's living standards were lower than those in cities. Moreover, because loudspeakers only broadcast programmes selected by the Party, the system greatly facilitated the Party's ideological control over the population. Consequently, the sound broadcasting became an efficient medium to propagate the government policies and educate the public (Lin, 1993).

The major functions of the loudspeaker system were to provide farmers with political education, spread advanced agricultural practices and enrich cultural life in rural areas (Chang, 1989). Chu (1986) documented that loudspeakers were most helpful in reaching areas that could not be covered by radio signals and they could serve each commune and brigade specifically by catering to the needs of local communities. Farmers were briefed about latest weather forecast and organized for immediate group activity with the aid of loudspeakers. Moreover, loudspeakers had been used to broadcast agricultural training programmes to educate farmers on planting techniques, fertilizer selection and application of insecticides. To further ensure the dissemination of important government messages in the rural areas, broadcasting monitors or local Party cadres organized collective listening among farmers so that the government policies were discussed and understood (Chang, 1989).

As an efficient way of transmitting political, cultural and technological information to the illiterate mass in rural areas where there was a low penetration of radio sets, the wired system underwent rapid growth during the 1950s. According to Chang (1989), the wired broadcasting units increased from 835 in 1955 to 11,124 in 1959. With the government drive, the wired broadcasting system was further strengthened and expanded at the regional and local level in the 1960s and 1970s. Whereas only 90,500 rural loudspeakers were reported in 1955, the number reached 106 million by 1976 (Bishop, 1989; Chu, 1978). In the early 1980s, the wired broadcasting development reached its climax and a nationwide system was established. According to Chu (1986), China had established an extensive loudspeaker network reaching 88.4% of all communes, 93% of the production brigades, 88% of the production teams and 65% of the rural households by 1980.

In addition to building a nationwide sound broadcasting system, traditional modes of communication such as the Chinese theatre, story-telling, ballad singing and poetry reciting had been transformed into modern instruments to propagate government policies and diffuse agricultural innovations. The popularity and influence of the traditional entertainment channels among rural people were taken advantage of because they were especially effective in crossing social barriers (Liu, 1965).

Take for example, the politicizing of the traditional drama. Both organizational and thematic reforms were carried out to make the traditional entertainment serve the purpose of transforming the Chinese people. The themes of old plays that went against Communist doctrines such as superstition, ghosts and Buddhism were banned and new Communist plays using traditional techniques were created to reflect China's revolution, nation-building or class struggles. Besides ideological education, the new plays also played a role in promoting new social values and improving production. The CPC used the traditional form of entertainment to solicit public support for national campaigns. Scores of dramas were created to publicize the benefits of the Marriage Law, educate rural people about sanitation and promote the advantages of mutual aid and cooperation through forming agricultural cooperatives (Liu, 1965).

Modern drama troupes were formed during the Great Leap Forward campaign. According to Liu (1965), there were 240,000 rural amateur troupes in China's one million villages in 1959, which broke down into one troupe for every four villages. This further substantiated the extensive as well as intensive political penetration into the rural regions. The dramas became a popular tool to aid agricultural production, promote new socialist values, and create a new Communist follower.

Due to a lack of recreational activities, going to movies was a popular pastime in rural areas. Chu (1986) reported a sprouting of commune and township cinemas in the late 1970s. The 90,000 mobile projection teams travelled among remote villages to show feature and educational films outdoors. Foot-pedalled generators were utilized in areas with no supply of electricity. Indigenous media such as wall posters, blackboards, bulletins, and mimeographed sheets served as alternative channels of communication for communes and brigades to spread information among farmers.

7.2.2 Rural Communication after 1978

The political, economic, and social reforms in the 1980s and 1990s have had a tremendous impact on China's communication system. Mass media are pragmatically used as a tool in nation building and fostering consensus for the reforms instead of as an agitator in class struggles and mass campaigns (Huang, 2003). The transformation in the role and function of mass media also affected rural communication.

Despite an overall significant increase in newspaper readership, and radio and television ownership throughout the 1980s and the 1990s, the expansion of mass media is mainly an urban phenomenon. The rural communication system remains relatively underdeveloped in terms of both the diffusion of technologies and the provision of media services, which has been a drag on the progress of rural modernization.

Concomitant with the rural reform in the late 1970s and early 1980s, there was a surge in rural newspapers because farmers were eager to get informed about production and government policies. The Association of Farmer Newspapers was established in 1984. According to *China Journalism Yearbook 1982*, China had a total of 476 national, provincial, local and specialized newspapers, among which 38 were related to agriculture with a total circulation of 9,260,000. However, the amount was still considered insignificant as compared to 857 million rural people at that time (one copy for every 92 farmers). The circulation of periodicals was also small in rural areas (155 out of 2,191 in 1980) and books for rural readers were in short supply (CJY, 1982).

It was not until the 1970s that the government started to emphasize the development of household radio sets. As a result, the penetration of radio sets paralleled the expansion of loudspeakers during this period though the increase rate of the former was much lower than that of the latter. Lull (1991) reviewed that household radio ownership in cities reached 32.3% while the figure in rural areas was 18.2% by 1982. The combination of the wireless and wired broadcasting systems enabled sound broadcasting to reach almost everybody in the country. By the end of the 1980s, almost every urban household had a radio set and every rural home was equipped with a loudspeaker.

The 1990s witnessed the decline of the wired broadcasting system. Owing to the rapid penetration of radio and television, loudspeakers were no longer the primary channels by which farmers remained connected with the outside. In 1990, the number of loudspeakers decreased to 82 million, penetrating only 37% of the rural households. The wired broadcasting was used less frequently at better-off villages where radio and television sets could be afforded. Furthermore, the implementation of the household contract responsibility system provided fewer opportunities for rural families to assemble, which contributed to the decline of the loudspeaker system (Lee, 1994). The film industry, likewise, dwindled because of the competition from television. There was a sharp fall of projection teams from the 1980s to the 1990s because of the high cost in film production, the entertainment alternatives provided by television and the extremely low fares charged to rural audiences (Lee, 1994).

China first introduced television in 1958. However, only in the early 1980s did the television industry start to develop into a full-fledged mass medium. The 1980s witnessed the booming of television not only in terms of its household penetration, but also the setup of new television stations and the production of programming content. According to the national survey conducted by China Central Television in 1987, 55% of the 1.07 billion (590 million) people were able to watch television. Of those television viewers, 370 million were urban residents (93% of the urban population) whereas 220 million lived in rural areas (33% of the rural population) (Lee, 1994).

Besides an enrichment of cultural life, television became a means of education for rural people. In 1979, television universities were established to provide correspondence courses for distance learning. As more and more farmers were able to afford television sets, television viewing quickly became an effective way of acquiring information and knowledge on agricultural technologies and general science (Chu, 1986).

The launch of communication satellites in 1984 and 1986 signified a breakthrough in China's telecommunication system. As there was an urgent social demand for trained manpower, which could not be satisfied through conventional education, a satellite television education system was established to increase 1.2 billion people's access to higher education in 1986. The satellite technology had a tremendous impact on mass education in a vast country like China. As mountains and deserts take up 80% of China's territory, those remote areas where it was hard to build cables, microwave trunks or television stations could be reached by satellite signals (Gao, 1991).

In the late 1980s, while television relay stations were established in big cities and large communities, TV Receiving Only (TVRO) earth stations were built in small towns and villages to receive programmes broadcast over the two available satellite television channels. Some TVRO earth stations were equipped with videocassette recorders to record programmes for later viewing. The satellite television education programmes offered courses and lectures on agriculture, and science and technology, which became a chief way of learning for people in mountainous regions and remote islands (Gao, 1991).

In addition to the electronic media, face-to-face communication also played a major role in information dissemination and knowledge sharing among farmers. Interpersonal communication was often involved in short-term training courses, visits paid by agricultural experts and professionals, demonstrating households, science-technology vans, science popularization fairs and horse-back information teams (Chen, 1991).

Most of these agricultural activities were organized and managed by the government. The Ministry of Agriculture, Animal, Husbandry, and Fishery and the Bureaus of Agriculture at the provincial and county levels were responsible for carrying out agricultural policies, while the Chinese Association of Agricultural Science usually sponsored scientific research, disseminated agricultural information among farmers, and provided them with technical training and advice (Chu, 1986).

Training courses were organized to help farmers improve their general knowledge as well as acquire specific skills on farming and breeding. The lectures were usually given during non-farming seasons and the content was adapted to the local needs and conditions. Moreover, Party leaders and agricultural experts often visited the rural households or places where innovation adoptions yielded good results. The personal contact could facilitate information exchange between experts and farmers (Chu, 1986).

Demonstration through experimental fields and exemplary households was one of the effective ways of diffusing agricultural innovations. Agricultural technicians devoted their efforts to one experimental field and demonstrated the results to farmers nearby. Convinced by the visible benefits of new methods of breeding crops or applying fertilizers, farmers might quickly adopt the innovations. Exemplary households referred to those who grasped new farming techniques and their pieces of land produced higher crop yields. They often shared the information and lessons learned with people in the vicinity. This kind of information sharing among villagers could be under the initiatives of either local governments or rural individuals (Chen, 1991; Chu, 1986).

Due to the shortage of print media in rural China, science-technology vans, science popularization fairs and horse-back information teams became chief channels for farmers to get access to educational and technological reading materials. At the science popularization fairs, farmers could consult agricultural experts on farming issues. Most often, science education films or slides were shown and short lectures were given to spread agricultural knowledge on these occasions.

7.2.3 The Contemporary Landscape

With the advent of the 21st century, TV viewing, radio listening and word of mouth are still major communication channels by which the majority of rural people get informed. Surveys indicate that farmers averagely spend about two to three hours watching television per day and 95% of their information is obtained from radio and television broadcasts (H. Zhang, 2005). As radio and television programmes are popular and easy to understand, they have become a significant part of rural people's daily life. Farmers often listen to radio or watch television to get updated information on government policies, learn of knowledge and technology, and thereby increase their personal capabilities. Moreover, the traditional broadcast media are widely used in rural areas as a major pastime due to a lack of recreational activities (H. Zhang, 2005).

After nearly five decades of development of the national broadcasting network, the Chinese government is still confronted with the formidable task of extending radio and television coverage to the vast rural areas. With the penetration of radio and television reaching the saturation points in urban centres and densely-populated rural areas in recent years, the government starts to target communities at remote and mountainous regions. The State Council has incorporated the "Extending Radio and TV Signals to Every Village" project into its development strategies since 1998. During the past seven years (1998–2005), the government has successfully extended broadcast signals to 117,000 electrified villages with an accumulative investment of RMB1.76 billion (U.S.$220 million), which has enabled over 70 million rural people to receive radio and television broadcasts (Yu, 2005). The survey conducted by the National Bureau of Statistics shows that rural ownership of colour television sets has increased to 67.8 per hundred households, whereas the equivalent number for black and white television sets is 42.8. The popularization rates of radio and television in rural areas stood at 92.4% and 94.3% respectively in 2004 (H. Zhang, 2005).

Despite the persistent efforts, there are still around 80 to 90 million rural people who are totally bypassed by the broadcast media. Moreover, the qualities of rural broadcast services are far from being satisfactory. A majority of the existing rural viewers (about 70%) can only receive two to three television channels via wireless transmission. Due to the low transmission power of relay stations and unfavourable weather and terrain conditions many areas have a poor reception of television signals (T. Wang, 2005).

Moreover, there is a severe shortage of television or radio programmes targeting rural audiences. Under the permission of the State Administration of Radio, Film and Television (SARFT), only ten radio frequencies and five television channels are related to agriculture and rural life, taking up about 0.42% (out of 2,371) and 0.4% (out of 1,254) of the total number of frequencies and channels in the country (H. Zhang, 2005). With the commercialization of the communication sector, media operators tend to focus their attention on the urban elites and their consumption patterns, whereas rural viewers are to a large extent marginalized even though they greatly outnumber their urban counterparts.

With an aim to build a public broadcasting system and achieve universal access to electronic broadcast media, the "2005 No.1 Circular" of the CPC's central committee required that first priority should be given to improving the quality of broadcast service with respect to rural issues. Accordingly, the SARFT set out a concrete objective of covering 93,926 villages with broadcast signals and providing the majority of rural audiences with access to at least eight television channels in 2005 (T. Wang, 2005). Despite the huge costs involved in diffusing media technologies in rural areas[15], the Chinese government is determined to employ various communication technologies, be it wireless, cable or microwave, to ultimately connect every rural household with broadcast signals.

15 It is estimated that a single percentage point expansion of the penetration rate may cost between RMB 3 and 4 billion (U.S.$360 to 480 million) (T. Wang, 2005).

7.3 The Development of the Internet

Driven by the liberalized economic environment and the advent of ICTs, the telecommunication industry has undergone rapid growth over the last decade. China has grown into the world's largest telecommunication market in dollar terms and one of the leading manufacturers and exporters of telecommunication equipments. According to the Ministry of Information Industry (MII), China's IT industry had an annual growth of over 20% between 1998 and 2002, which was three times the growth rate of the economy. The percentage of added value of the telecommunication industry in GDP rose from 1.5% to 2.7%. In 2003, China's telecommunication service revenue totalled RMB 515 billion (U.S.$ 64 billion) ("Zhongguo Zheng", 2004).

In terms of its capacity to participate in ICT development and the potential to exploit technological innovations, China has been ranked as a country with middle level development on the basis of three major E-readiness assessments (TAI, NRI and DAI) (ITU, 2003; UNDP, 2001a; WEF, 2003).[16] Teledensity (the sum of fixed-line and mobile

16 In the Human Development Report 2001, altogether 72 Countries have been categorized into four groups according to their TAI values due to the availability and quality of corresponding data: Leaders (TAI above 0.5), potential leaders (0.35-0.49), dynamic adopters (0.20-0.34) and marginalized (below 0.20). China comes under the category of dynamic adopters with a TAI value of 0.299, ranking 45 (UNDP, 2001a).

Among the 102 countries included in the Global Information Technology Report 2003-2004, China is in 51st place, with relatively higher ranking for the Readiness component index (30), the Usage component index (43), but more disappointing performance in terms of the Environment component index (63) (WEF, 2003).

According to *the World Telecommunication Report 2003*, the DAI has been designed in a much simpler way than other indexes so as to cover the widest number of countries and enhance transparency. In the latest report, 178 economies have been categorized into four groups based on the level of ICT access: high access, upper access, middle access and low access. The top five countries

users per 100 inhabitants) of China was 0.6 in 1990, and reached 42 in early 2004. By the end of April 2004, China had a total of 581 million telephone subscribers with 285 million fixed and 296 million cellular, whereas the total number of telephone lines stood only at 44.3 million in 1995 (Letcher, 2004).

Currently, China's telecommunication market is dominated by four major operators: China Telecom, China Netcom, China Mobile, and China Unicom. Following China's accession into the WTO in November 2001, China Telecom was split up into China Telecom and China Netcom, with the former catering to the south and west of the country and the latter having large assets in northern China. China Mobile is the world's largest mobile operator by number of subscribers, which provides services to over 150 million users.

On April 20, 1994, China opened a 64K international dedicated circuit to the Internet through Sprint Co. of the United States, realizing a full-function linkage. Since then China has been officially recognized as a country accessible to the Internet (CNNIC, 2003).[17] The rapid development of the Internet has widened Chinese people's access to information. It has revolutionized the ways Chinese communicate and interact, enabling them to achieve many goals online, ranging from education, meeting people, playing video games to shopping.

In recent years the central government has invested heavily in telecommunication infrastructures to encourage and support deployment of the Internet. China had laid down more than one million kilometres of optical fibres by 2000 (Long, 2001). With these visible efforts, the Internet in China is undergoing phenomenal growth. After modest beginnings in the mid-1990s, the number of Internet users increased dramatically in 1998 and 1999, reaching 8.9 million users by

---

are Sweden, Denmark, Iceland, Korea and Norway while the lowest is Niger, with China in the middle access group (ITU，2003).

17 The China Internet Network Information Centre (CNNIC) is a semi-official non-profit organization that is administered by the Chinese Academy of Sciences and involves itself in Internet-related matters, which are within the purview of the MII.

the end of 1999. China now boasts the world's second largest population of Internet users [18], the number of which increased to 111 million people by December 31, 2005, a big jump from 0.62 million users in 1997 (CNNIC, 1997; CNNIC, 2006) (see Figure 5). There are 49.5 million computer hosts, about 2,592,410 domain names and 694,200 websites in China (CNNIC, 2006).

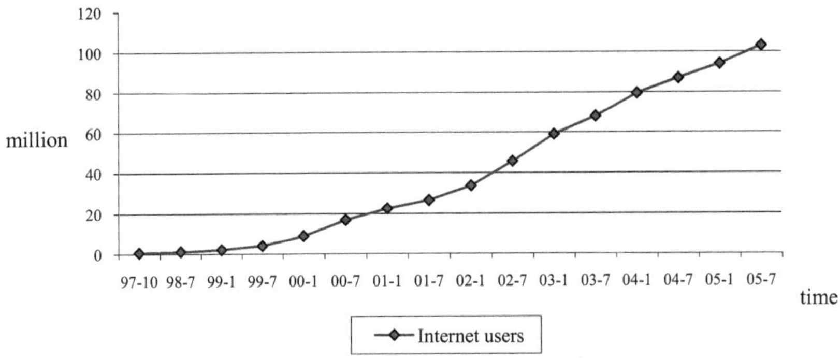

Figure 5. Growth of Internet users in China (1997–2005).[19]

Although the Internet use is expanding much faster than expected, net surfers account for only a very tiny fraction of the national population (about 8.6%), with the majority of the users between 18 and 35 years old. As in other countries, most Chinese net surfers use the Internet from home (70.5%), while others log online at work (37.6%) or at Net cafes (27%). Broadband and dial-up are the major modes of accessing the Internet (CNNIC, 2006).

Despite its phenomenal growth, Internet penetration is basically an urban phenomenon that excludes the majority living in rural and remote regions. The total number of Internet users in rural China is about 1.5 million, accounting for 1.4% of the total Internet users and

18 CNNIC defines the Internet user as a Chinese citizen who uses the Internet at least one hour a week.

19 From the China Internet Network Information Centre.

0.12% of the population (CNNIC, 2006) (see Table 3). A major impediment to rural Internet usage is the inadequate telecommunication infrastructure. Rural and remote areas are much lagging behind in terms of the diffusion of both traditional and new media technologies. The urban household telephone penetration rate was 1.61% higher than that of rural areas in 1990, while the gap rose to 25.96% in 2002. Urban residents can receive 30 to 40 television channels, while rural people can only enjoy two or three (Z. Zhang, 2005). A great number of rural people suffer from a short supply of electricity and substandard power networks. The 2002 statistics show about 30 million people in 1,061 townships (mainly in remote rural areas) did not have access to electricity (Wacker, 2000).

Table 3
Farmer Internet Users

Survey	Users (Farmers)	Percentage in total users
01/2006	1,554,000	1.4%
07/2005	1,236,000	1.2%
01/2005	940,000	1.0%
07/2004	1,044,000	1.2%
01/2004	636,000	0.8%
07/ 2003	612,000	0.9%
01/ 2003	591,000	1.0%
07/ 2002	503,800	1.1%
01/ 2002	438,100	1.3%
07/ 2001	318,000	1.2%
01/ 2001	171,000	0.76%

Note. From the China Internet Network Information Centre.

The widening urban-rural income gap also greatly constrains rural people's consumption of high-tech products. Personal computers, which are necessities for many urban residents, remain an extravagant purchase for rural people. The urban computer ownership reached 20.63 per hundred people by the end of 2000, whereas the figure was almost zero with the rural population (Z. Zhang, 2005). Lack of farms, absence of required funds and poor educational background make Chinese farmers differ from their Western counterparts. Owing to low educational levels and lack of sufficient channels to get information, words like “Internet” and “surfing” are still foreign to many rural residents. The unequal access to information has resulted in a huge digital divide between urban and rural people.

Despite its low penetration in the vast rural areas, the Internet is generally perceived as a new engine for rural empowerment and quite a number of experimental projects have been launched in recent years to test such an assumption. Chapter 8 to Chapter 12 will present the findings on Internet diffusion and usage at five rural places.

8. Yellow Sheep River Town

8.1 The Profile

Yellow Sheep River is an impoverished mountainous town under Gulang County, bordering the desolate Gobi Desert. As one of the eighteen drought counties in Gansu Province, Gulang is among the key areas for national poverty alleviation. In Yellow Sheep River Town, about 23,000 rural people dwell in 81 settlements within 175 square kilometres, most of which are located in hard-to-reach mountain areas.

The natural environments do not favour the local residents. With an annual rainfall of 300 mm, the yearly harvest is by and large determined by the weather. Many residents living in the mountains have to trek over 20 km everyday to fetch drinking water. Up till now, some households still keep the habit of storing rain or snow for drinking water. In a lean year, there may be no yields at all.

Because of the formidable natural conditions and a dry climate, the local agriculture has been dominated by the self-sufficient economy, with traditional farming and livestock rearing as the chief means of making a living. The harvested crops go first to the daily food provision. The remains, if there are any, are sold in the local market where a barter trade still goes on.

As 98.7% of the town people are engaged in agricultural production, the per capita annual income stood at less than U.S.$140 in 2004 ("Gulang Gaikuang", 2005). There is a high rate of school drop-outs as youngsters have to work to supplement the limited family income. The migrant workers on average bring back a yearly income ranging from U.S.$366 to U.S.$732, which helps the households make ends meet in times of a drought. Consequently, the heavy physical labour involved in land cultivation is more often shouldered by the old and the female at home.

People's life styles and social norms remain unchanged. TV and word of mouth are two major ways for the local farmers to get informed and entertained. Thanks to the project of extending radio and TV coverage to every village carried out by the Chinese government, most of the villages are now reached by TV signals. However, the high subscription fee plus the limited choice of programmes keep the majority of households from accessing the service provided by the county cable TV station. Alternatively, satellite dishes can be found at most of the homes of the informants despite the fact that they are outlawed by the state regulations. X. Wang (2005) estimated that 80% of the local households could view approximately ten channels via the satellite dish.

Another way for the rural residents to learn of the outside world is through the migrant "farmer workers", who constantly travel between big cities and their hometown. The great social mobility resulting from China's market reform has given rise to a new social class of migrant workers, whose social status is rated higher than that of farmers (Lu, 2004). These young people bring back not only considerable earnings but also vivid descriptions about life in the cities, intermediated by their own perceptions.

The wired loudspeakers and the wireless radio sets have given way to TV. The subscription to newspapers and magazines is extremely rare among the farmers due to their poor economic status. Even though the whole village may subscribe collectively to some local newspapers, they are for the exclusive use of the village heads and therefore not within the reach of the common people (X. Wang, 2005). According to statistics provided by the county government, there were six fixed-line telephone users per hundred, while the mobile phone subscription remained as low as 1.5 per hundred in 2002.

The majority of the adult farmers have not completed the nine years of compulsory education and the elderly remain illiterate or semi-illiterate. The Yellow Sheep River vocational middle school is the only one of its kind in town, which enrols children living within a distance of 30 km. It is in such social and economic contexts that the Internet was introduced to this town in 2000, starting first with the vocational school.

8.2 The Change Agency

The Internet came to this rural town as an experiment known as the "Town and Talent Project" initiated by Wen Sayling[20], who was the late vice chairman of a Taiwanese company. Wen decided to invest in the local Internet project after he heard about the local people's plight of impoverishment from one of his employees, who had spent a year in the town's vocational school as a volunteer teacher in 1999.

The idea of the "Town and Talent Project" was nurtured during a series of charitable initiatives to enrich the educational resources of the school. The Taiwanese company, which was specialized in software development, donated eleven second-hand computers and 1,500 books collected from its employees to the school in July 2000. In the same year, the first computerised classroom was set up with another eleven computers and network facilities purchased by the school on credit.

20 Wen believed the Internet had demonstrated beyond all doubt that it could change the course of people's lives. Only by way of the Internet could one eliminate the digital divide and assist underdeveloped areas in bidding farewell to poverty. Hence, Wen and his followers established Town and Talent Technologies with a 50-million-U.S.dollar investment in 2002. Wen expected that the modernization of rural villages could be accomplished within 10 years through the use and popularization of computers and network technology. But in December 2003, Wen suffered a stroke and died in Taiwan at the age of 53. Now it is Kenny Lin, Wen's friend and college classmate, who is running Town and Technologies and trying to salvage Wen's dream by setting up small digital centres in western remote regions.

The company also invited two teachers from the school to receive computer training at its Tianjin headquarters. With technical support from the company, the two teachers helped the school set up its own network, which was connected with the Internet in October 2000. Closer ties were fostered between the school and the company as the students and teachers exchanged visits with the employees.

Triggered by the rapid changes taking place in the school as well as motivated by the humanitarian concern for the deprived rural people, Wen hoped that the Internet could be deployed as an efficient tool to narrow the digital divide and assist underdeveloped areas to eliminate poverty. With an investment of U.S.$50 million, Wen set up the Town and Talent Technologies Cooperation in 2002 with the ambitious target of transforming the impoverished local communities from an agricultural society into an information society through the popularization of computers and network technology and fostering knowledge-based economic development there.

The planners envisaged seven steps to achieve the ultimate goal of elevating the level of knowledge, income, and living standard of China's west to that of the eastern coastal areas: schools leading local economic development, expanding e-business, improving agricultural and pastoral economy, establishing Internet villages, encouraging relocation from rural settlements to the Internet villages; developing the service industry; and enacting a system of remote employment and service.

As every rural town had at least one secondary school with relatively better educated teachers, schools were perceived as leaders in promoting the use of new ICTs in rural areas. Therefore, the project chose the vocational school in Yellow Sheep River to lead the pioneering experiment. It was hoped the model would be replicated in other parts of western China through the involvement of 1,000 secondary schools and the establishment of 100 branch offices in 100 cities. The Town and Talent Technologies would also employ and train 10,000 software engineering technicians.

The project planners expected that the modernization of rural China could be achieved within ten years by building Internet villages and developing knowledge-based economies. This pioneering project thus provided us an opportunity to observe and document the changes that the Internet had brought to the vocational school and the rural life in Yellow Sheep River.

8.3 The Internet and Education

As the first school involved in the project, the vocational school at Yellow Sheep River had received 140 donated computers since 2000. With the setting up of two computer classrooms as well as the school website, teachers and students started to enjoy the benefits provided by the new technologies.

Teachers were among the first to adopt the computer skills and Internet applications. Due to their lack of experience with the technology, some elder teachers were scared about computers and worried that they could not learn the sophisticated skills. In order to help the staff get rid of the technophobia, two training courses were designed to cater to teachers with different levels of prior knowledge. The elementary courses focused on the basic operation skills of Microsoft office systems, whereas advanced courses taught the designing of course software and web pages. Within several months, 90% of the staff grasped the basic E-literacy, which paved the way for the widespread employment of the Internet technology in school management as well as daily teaching and learning.

With the broadened channels of information and communication, teachers were able to access the latest educational materials and therefore became more receptive to modern education ideas. While the old fashioned class preparation was on the basis of textbooks and references provided by the Ministry of Education, teachers could download the course software at the websites of elite schools in Beijing and Shanghai, which functioned as templates for them to organize the lectures.

The online education resources contributed to fostering academic atmosphere on campus, which greatly motivated the teachers to improve their teaching. On a voluntary basis, teachers were engaged in discussions on how to select and compile the online materials so as to enrich the content as well as improve the teaching methods. Students' academic performances were recorded and analyzed to test the desirability of new education practices and monitor their learning progress. Under the collective efforts, an E-database of multiple disciplines was set up by cataloguing large amounts of teaching software and audio and visual materials downloaded from the Internet.

Convinced of the perceived efficacy of the Internet in education, teachers' attitudes towards the technology underwent sweeping changes. The purchase of a networked computer at home became enviable and the heated online discussion often provided discussion topics for everyday life.

In addition to the wide adoption of the Internet technology among teachers, students started to change the traditional ways of learning with the command of computer and Internet access skills. Previously, due to a shortage of reference materials in school, knowledge acquisition was confined to textbooks and lecture notes. Computer skills and Internet applications enabled the students to develop an active and self-paced learning and draw on the abundant information stored over the Internet.

The Town and Talent Technologies also installed self-designed software in the donated computers, which provided users with training on three key applications – e-mail, Internet surfing and English learning. Under teachers' guidance, students could practice their computer skills by following the software instructions. In their computer classes, students also participated in the trial run of the Entrance Exam One-Step Programmes developed to let students gain experience with the college entrance exams through simulated tests.

Despite the easy access, the schoolers were forbidden to log on the Internet directly as teachers worried that they might access undesirable information online. Instead, teachers would select news and information that they perceived to be useful to the young and put them on the server of the school network. Students were allowed to browse the local network freely, where only carefully selected online information was on display.

The cyber world aroused the curiosity of the young and greatly broadened their horizon. The web-based content served as an aid and supplement to classroom teaching. Students learned to search information on the local network as well as the E-database to seek solutions by themselves when they stumbled over problems that were not clearly explained in class.

Moreover, in a life environment where limited TV access and word of mouth served as the chief means of getting informed, the Internet enlarged the visions of the schoolers trapped by remoteness and isolation. The young informants reported that they found an outside world more extensive and more colourful than the bleak view of Yellow Sheep River. In their eyes, the Internet could accomplish such feats as shopping, job hunting, contacting friends and information searching. They also sought entertainment by listening to popular music and songs downloaded from the web.

A 17-year-old girl named Zhao Mei said that she was among the first batch of students who had the opportunities of attending computer classes in 2001. Previously she only knew a computer was somewhat like a TV set. With the help of teachers, she grasped the skills of typing and surfing the Internet. Zhao Mei found a bigger world outside Yellow Sheep River. She often told her parents about the news she read from the websites, which kept them also informed of what was taking place in other parts of the country or even the world. Zhao realized that she did not have to get married immediately after finishing middle school. Instead, she could seek further education and find a job in the cities.

Even though the students were not granted direct access to the Internet, the school websites provided them with a good platform to practice the network skills they learned in class. Nearly a hundred web design projects were kicked off to introduce the town and the school to the outside. Teachers and students wrote articles and took pictures of local scenery, customs, and specialties. The improved school websites allowed more people to learn of the school life as well as the local resources.

Teachers and students started to set foot outside Yellow Sheep River to attend webpage design competition and SET (namely software, English, typing) skills contest in Beijing and Tianjin, together with teams from other Town and Talent member schools. More and more educators from universities in Beijing, Hong Kong and Taiwan volunteered to hold seminars jointly with the rural school.

The implementation of the project brought fame as well as ample opportunities for the vocational school. The education reform sparked off by the application of the Internet to teaching was rewarded by the rapid progress that students made. There was a sharp increase in enrolment and more and more students were admitted to high schools after the self-administered exam software was introduced to the curriculum (see Table 4). The lowest drop-out rate in history was achieved and girls would not think of getting married after graduation as the only life choice. They were anticipating a better future by seeking higher education.

8.4 E-Commerce, Manager Hu and Social Support

At Yellow Sheep River Town, the Internet technology was diffused mainly through two groups of people: students and teachers on campus, and scattered farming households off campus (see Figure 6). Concomitant with the rapid diffusion of the Internet among teachers and students, efforts were made to extend the benefits of the technology to the local farmers. Driven by the objective of school leading the local economic development, the "Town and Talent Project" established an E-commerce centre on campus with the hope of fostering E-readiness among farmers and helping them change the traditional way of farming and selling.

Table 4
Students' Enhanced Academic Performance

The number of enrolment					
Year	1997	1999	2002	2003	2004
Number	653	698	947	1,090	1,283
The rate of students with a passing grade (%)					
Year	1998	2001	2002	2003	2004
Junior 1	6	12	15.8	18	22
Junior 2	7.5	10.2	14	16.3	18
Junior 3	13	22	24	25	30
The rate of students with an excellent grade (%)					
Year	1998	2001	2002	2003	2004
Junior 1	72	71.6	72	80.3	82
Junior 2	69	69	69.2	71	71.4
Junior 3	77	80	80	85	85.7

Note. From [The E-commerce achieved by Yellow Sheep River vocational school], by W. Hu, 2005, unpublished raw data.

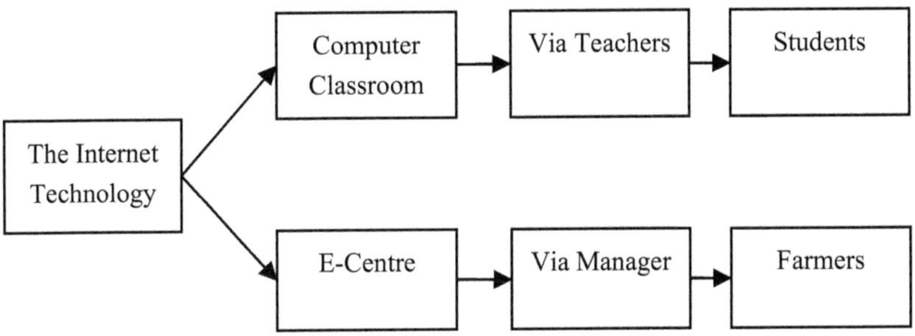

Figure 6. The Internet diffusion pattern at Yellow Sheep River Town.

The E-commerce centre, equipped with 25 computers, a printer, a digital camera and a fax machine, was registered and financed as a business under the "Town and Talent Project". The Town and Talent

Technologies supervised the daily operation of the centre and offered technical support whenever necessary.

The local farmers could learn how to use a computer through the E-commerce centre, outside of which information on farming skills, land policies and prices of seeds and fertilizers were posted on the walls when the weather was fine. People would usually stop and read the information when they came for the regular town fairs. The off seasons for farming were the busiest time for the centre. Since its opening in 2000, it had organized two dozen Internet training courses for farmers as well as 36 training sessions tailored to the special needs of the village chiefs, entrepreneurial farmers, and students from the schools nearby. The most-often taught courses included basic knowledge about computer hardware, frequently used office software and Internet surfing.

The centre was frequented by both teachers from the school and local farmers. It did not charge teachers for their Internet use because the Internet service was considered part of the teaching facilities. Farmers were charged RMB 2 per hour to use the service provided by the centre. The running costs of the centre such as Internet access, electricity, phone calls, and hardware replacement, were covered by the tuition for the training sessions, fees from daily users and the 1% commission charged for every transaction completed via the E-commerce service.

The chief administrator of the centre was Hu Wanlong, a history teacher of the vocational school. As a part time administrator and manager of the E-centre, Hu was responsible for searching information for farmers, offering technical support, maintaining the hardware and keeping the centre financially viable. In addition to managing the centre, Hu also served as an informal advisor to the farmers on issues ranging from a new type of tractor they saw in neighbouring villages to the latest government policies on land and farming.

Despite the training, few local farmers could actually operate the computers and access the Internet by themselves. As a result, their use of the Internet was most often mediated by Hu, who became a central figure because most social changes that the town experienced mainly revolved around the E-commerce centre under his management.

Hu's reddish cheeks were no different from those of other dwellers at Yellow Sheep River Town, a look typical of people living in high land regions as they were wizened by harsh winds and mountaintop sunrays. A graduate from the vocational school himself, Hu came back to lecture in his hometown after completing college education in Lanzhou, the capital city of Gansu Province. Hu said with pride that he was the first person in town with some computer know-how. It was in March 1999 when the town welcomed its first computer, a gift sent to the local government by an educational institution in Lanzhou. Since nobody there had the technical skills to operate the machine, which was installed with DOS system at that time, the machine was given to the vocational school. Hu started to use the computer in preparing examination papers and office documents.

The experiences accumulated over the early years as well as strong personal interests enabled Hu to become a computer expert among his peers. Hu's technical competence and seniority in school won him the trust of the principal as well as other teaching staff, who sent him to receive advanced training on software and web page design in Tianjin. Equipped with a refreshed knowledge about Internet applications, Hu demonstrated his capabilities through building up the local network as well as the school website. During the initial stage of cooperation between the school and the donating company, Hu also served as a chief coordinator to facilitate the exchange of information between the two organizations.

In the diffusion process of the Internet at Yellow Sheep River, Hu became both the obligee as well as the obliger. His readiness and wittedness in quickly adopting the Internet helped him gain high social prestige among the fellow villagers. As soon as he acquired the technical competence with the Internet, Hu used the Internet as a means to empower the helpless people around him. Hu and the early adopters formed the central links of a web, through which the awareness of Internet applications was slowly diffused to the scattered rural households.

When talking about his experience with the Internet, Hu felt quite contented with his achievements. Unwittingly, he seemed to feel that he had accomplished more as the manager of the E-centre than what he had done as a full-time school teacher. As far as individual farmers were concerned, those who benefited most from the Internet services were either Hu's acquaintances or villagers who were aware of the existence of the E-centre. The Internet proved to be an efficient means of empowering the villagers, especially those in an economically disadvantaged position and therefore in greater need of support. To these people, Hu functioned as a bridge between the real world and the virtual world.

Through Hu's efforts, the Internet was utilized as an efficient tool for the rural individuals to strive for citizen rights and protect their own interests. One of the boys in town was handicapped due to a bad surgical operation in a city hospital. The family turned to Hu for help. By surfing on the Internet for information regarding medical compensations, the victim's family and Hu learned of the legal process of suing the hospitals as well as some of the supporting cases. The family finally won the case and used the compensation to help the boy recover through a second operation.

The boy would not have had his normal and healthy life back had it not been for Hu and the E-centre. In the remote mountainous areas, rural people often did not know how to redress grievances due to their ignorance of legal regulations. The situations became even worse when the wrongs were inflicted upon them by the privileged few working at government or public institutions. Consequently, the victims often chose to remain silent because they had neither the knowledge nor the money for a law suit, which might take years to come to a conclusion. However, the large pool of specialized information available online functioned as valuable resources to be drawn upon by the deprived segment at virtually no cost, which would greatly enhance their personal capabilities of dealing with social injustices and keeping their own destinies under control.

Hu also helped one of the rural households seek social and financial support via the Internet. Li Xingjiang, a primary school teacher in a neighbouring town, had a daughter who had been diagnosed as having a brain tumour at the age of 12. The parents became so desperate that they would rather sell their kidneys to raise money for their daughter to undergo a surgery. Deeply touched by the plight of the family, Hu posted the girl's story on the school website and called for donations. He also made CDs describing the girl's illness so as to solicit social help.

The event soon attracted the attention from the Town and Talent Technologies as well as the provincial government. The director of the Department of Health of Gansu Province inquired about the girl's illness and arranged for her to enter the best hospital in the province. A medical consultation was subsequently held among brain experts to come up with the safest way of treatment. The girl finally had a timely surgery with the donations to the account listed by her father on the school website.

Li Xingjiang gradually adopted the Internet during the process of seeking treatment for his daughter. Under Hu's instructions, Li learned to search information online about her daughter's illness and exchange e-mails with medical experts and friends who were concerned for the young girl. The personal experience gave him an opportunity to fully understand and accept the Internet technology. Previously, most of the villagers tended to associate the Internet with computer games and did not believe in Li when he mentioned about his attempt to solicit help from the Internet. However, the sceptics started to get convinced about the efficacy of the Internet after noting the emotional and financial support Li's family received. Hu and the E-centre made the advantages of the Internet tangible and convincing not only to Li and his family members but also to the neighbouring villagers. The local people began to realize the Internet was of more use than they could possibly imagine before.

In addition to socially empowering the rural individuals, the E-commerce centre devoted itself to improving the traditional way of cultivation and selling. The farmers started to accept the concept of scientific farming after exposing themselves to various agricultural websites at the training sessions organized by the E-centre. Previously, few farmers knew about the techniques of fertilizing their lands. Most often, the amount of fertilizer to be used was not determined by the relative proportion but by the annual incomes. The more money made, more fertilizer would be used. The agricultural websites offered useful tips to the villagers as to how to scientifically fertilize the land and get rid of weeds and insects at appropriate time. Furthermore, many online dealings, such as the selling of green peas, yams and celeries, were accomplished by posting information about the local farm produces online, the total transactions of which amounted to RMB1.3 million within five years (W. Hu, personal communication, April 8, 2005).

Despite the initial success of online dealings, there was little growth in the volume of transactions as time went by. According to Hu, only about a dozen farmers visited the E-centre within a year. The town's website still advertised locally grown mushrooms, but Hu said they were not really selling anymore. The only thing that still attracted customers was a ginseng-like dried root used in traditional medicine. But it grew in tiny quantities and was hard to propagate on a large scale. As aforementioned, the few early adopters tended to be Hu's acquaintances and Wu Zhenhou was one of them.

Wu Zhenzhou, an old acquaintance of Hu, ran a clinic at a neighbouring town located in Tianzhu Tibetan County. All the villagers would go to his clinic to seek treatment when they fell ill. Wu used to sell medical herbs collected from the mountains to some middlemen, who often cheated him by not paying. With the help of Hu, Wu managed to set up contacts with a few pharmaceutical companies in the province through the Internet and sold his herbs at a higher price.

Wu said that the majority of the farmers in his village did not know anything about the Internet. The households in mountainous areas were so scattered that even people in the same village were not aware of the happenings in their neighbours' houses. The chief modes of getting informed were through gossip or the fixed line pay phone at the central village store. Even though farmers had mountain produces to sell, buyers were non-existent. Therefore, they were often taken advantage of by the middlemen, or door-to-door traders, who either offered an extremely low price or cheated them by not paying at all.

Wu came to accept and understand the functions of the "electric Net", a term he used to refer to the Internet, through his experiences of selling via the web with the help of Hu. In his perception, the Internet was an efficient tool to increase income, exchange information and establish contacts. Wu often visited the E-centre when he went to the Yellow Sheep River Town. Sometimes he asked Hu to search price information on medicine so that he could get a general idea when making purchases for his clinic. Upon Wu's request, Hu also searched new books on pharmacy for him to buy.

Wu felt very proud that he was the first person who knew the Internet in his village. He said he would not have possibly appreciated the benefits brought about by the new technology if it were not for Hu. Wu reported that other villagers could not be so fortunate simply because they had no other ways of getting to know the Internet. He became more popular than before in the neighbourhood since many villagers began to send mountain produces to his house with the hope of securing a better offer.

Publicity over the Internet enabled the town to attract enough investment to build a potato processing plant. Potatoes, a major local produce, were sold at a low price because of their high yield. The new potato processing plant allowed the farmers to raise their income by turning potatoes into fried chips, which were sold at a higher price on the market.

The E-commerce centre also managed to provide farmers with job opportunities through the business contact established between the Town and Talent Technologies and a shoe-making factory in the coastal Guangdong Province in south China. All the would-be assembly line workers underwent training at Hu's centre before setting off. They were taught to send and receive e-mails and chat online so that they could still keep in touch with their family members while they were away from home.

It is obvious that Hu played a pivotal role in promoting the use of Internet among the local farmers, who had difficulty in using the

Internet for practical purposes due to the lack of general knowledge and computer skills. As an innovator, Hu had not only helped the farmers use the Internet to solve their problems, but also shaped their perceptions of the Internet in terms of what the Internet meant to their life and what it could accomplish for them. Without Hu, the digital divide would have continued to exist even though the farmers had been provided with the Internet access.

8.5 The Experiences of Yellow Sheep River and their Implications

Five years have lapsed since the Internet was introduced to Yellow Sheep River, which now claims to be the first Internet town in China. The new communication technology has brought fame and glory to this remote western town. Only by seeing students and farmers clicking mouses do we start to believe that computers and the Internet, the icons of the information society, may have bearing on this underdeveloped region in China.

The vocational school becomes a model IT training base in Wuwei City. Almost every year school teachers from twelve western provinces come here to receive computer training. With the technical support provided by the school, another E-literacy training centre has been set up at Danancun Village. Inspired by the noticeable effects achieved in the vocational middle school, six other middle schools in Gulang County joined the "Town and Talent Project" by replicating the development model of cooperation between rural schools and the private organization.

The use of networked computers for E-literacy training and stimulating economic development attracts attention home and abroad. Scores of overseas visitors come to see what has taken place in the small town. The vocational school receives wide coverage not only in the local newspapers, but also in national media such as China Central

Television (CCTV) and transnational media like Phoenix TV. The manager of the E-centre, Hu, becomes a local public figure after being featured in a CCTV programme.

The local government starts to react to the series of changes brought about by the Town and Talent Technologies. The officials from the provincial government as well as top leaders from the Ministry of Education paid visits to the school and proclaimed it as a role model in school computer education. They also endorsed the "Internet Village" strategy as the way to develop the local economy.

Subsequently, the town government has allowed the Town and Technologies to use a 43-acre lot for 50 years at no charge for setting up the Internet Village. The planners hope that once the farmers acquire digital readiness they will be relocated from their rural settlements to the Internet Village, where they can join the service industry and seek employment elsewhere. The international conventional centre serves as a central base for the villages to enter into the information society as well as a business platform between eastern and western China. The construction of the conventional centre has also helped to absorb the local surplus labour by creating several hundred jobs.

To grasp the development opportunities concomitant with the advent of the Internet technology, the local government also takes measures to improve the local infrastructures. The rugged road connecting the village to the main highway was widened and paved, with pine trees and willows planted along it. Mobile phone and broadband Internet services were introduced to the town in 2003. The county government has decided to enlarge the scale of the vocational middle school by constructing a new campus, which will greatly improve the teaching facilities and promote the educational level.

In Yellow Sheep River, the private investor company took the initiative to induce changes to the impoverished town by introducing digital technologies. The poorly educated mountainous dwellers and the conservative rural communities made the penetration of the Internet extremely difficult, if not completely impossible. As the vocational middle school had most of the better educated people in town, who could be more receptive to innovative ideas, the investor company naturally took the middle school as the central hub to extend the technology. Therefore, the "Town and Talent Project" was designed to let schools lead local economic development and the Internet was first introduced to teachers and students.

Despite the limited number of networked computers donated by the private company, the faculty and students as well as local farmers could use the Internet on a shared basis, which greatly reduced the cost of Internet access for poverty-stricken areas. The E-commerce centre managed by teacher Hu took on the task to promote E-literacy among farmers to change the traditional way of farming and selling farm produces. Most of the extension efforts revolved around providing computer training, information searching services and informal consultancy.

The information exchange involved in the innovation diffusion process mainly took the form of interpersonal communication. This happened not only between teachers and students on campus but also between Hu and people from scattered farming households. Despite the systematic and intensive training received in class, farmers still needed Hu to help with the Internet application due to low level of technical competence and lack of access after the training sessions. Consequently, E-centre and Hu played a pivotal role in disseminating advanced farming practices and innovative ways of selling among rural people. As an information gatekeeper, Hu facilitated farmers' Internet use and helped them cross the barrier of knowledge posed by the Internet. Without directly operating computers, farmers were still able to learn of advanced farming technologies and use the Internet as a tool for personal empowerment.

The experiment at Yellow Sheep River demonstrates that ready access to the Internet does not necessarily lead to direct usage of the technology by the farmers, who are yet to acquire the intellectual and technical competence needed for processing online information. To bridge the gap between the technology and farmers, online communication is likely to take multiple steps before information from the Internet can directly reach and benefit the rank and file of farmers. It is no surprise that rural schools, which serve as the knowledge centers in rural areas, were used in the project to cross the barriers between the farmers and the Internet.

Even though we are unable to obtain solid statistics to show the changes brought about by the Internet to the local economy and rural life in Yellow Sheep River, there are abundant evidences to show that the Internet has indeed benefited the farmers' life. The Internet has broadened the vision of the villagers trapped in their remoteness and isolation. More and more students have gone for higher education after the self-administered exam software was introduced to the curriculum. Girls no longer think of getting married after graduation as the only choice of life. They all have better hopes for life by seeking higher education. The Internet functions as a window not only for the villagers to look outside but also promote themselves to the rest of the world. Through its website, the small town has attracted external attention as well as more and more visitors.

As a Chinese saying goes, "A single spark can start a prairie fire." By October 2005 the Yellow Sheep River Model has been extended to 94 middle schools in 15 provinces, municipalities and autonomous regions, among which 87 are located in the less developed western China (see Table 5 and Figure 7).

Table 5
The "Town and Talent" Member Schools (by October 2005)

Provinces	The number of member schools	Location in the country
Gansu	12	Western
Shanxi	10	Western
Ningxia	30	Western
Qinghai	9	Western
Sichuan	12	Western
Chongqing	6	Western
Guizhou	1	Western
Tibet	5	Western

Guangxi	1	Western
Inner Mongolia	1	Western
Heilongjiang	1	Eastern
Jilin	1	Eastern
Zhejiang	1	Eastern
Beijing	2	Central
Hebei	2	Central

Note. From Huiyuan xuexiao. (2005). [Member schools]. Retrieved January 30, 2008, from <www.yellowsheepriver.com/schoolbase/index.php>.

Figure 7. The "Town and Talent" member schools over the country (by October 2005).[21]

21 The map shows 15 provinces, municipalities and autonomous regions where the "Town and Talent" member schools are located. The names of those provinces and regions without member schools are not indicated.

These schools are all linked to the central website of < www.yellowsheepriver.com > and thereby form a platform to share and exchange education resources. The school websites also play an important role in introducing their towns to the outside world. Therefore, the Internet is not only adopted in Yellow Sheep River but also at multiple sites of the impoverished western regions. Even though studies have not been conducted in other rural areas with Town and Talent member schools, the implications that the Internet has for the local education and economic development in Yellow Sheep River could be taken as an epitome to show what the Internet means to the life of the western farmers. However, the long-term impacts of Internet diffusion and use on rural development still take time to unfold.

9. Jinta County

9.1 The Profile

Jinta County is located in the northeastern part of Jiuquan City, Gansu Province. As one of the outposts on the ancient Silk Road, Jinta is a rural oasis bordering the Badain Jaran Desert of Inner Mongolia. In the county, about 138,000 rural people live in 123 villages under 13 townships within an area of 18,800 square kilometres, 1.4% of which is arable.

Jinta is rich in natural resources, especially in land, underground water, heat energy and minerals. In addition to the arable land, there are more than 1,000 square kilometres of wasteland suitable for cultivation and grazing. Since ancient time, Jinta has been famous for its irrigated agriculture thanks to two rivers running through the county and 14 reservoirs. Jinta also enjoys an annual sunshine period up to 3,193 hours and rainfall of 59.5mm. The sharp temperature difference, few precipitations as well as quick evaporation render Jinta an ideal base for breeding seeds and growing grain, cotton, corn, sugar beets, cumin, and forage, etc ("Jinta Gaikuang", 2005).

Due to the favourable natural conditions, Jinta is one of the national commodity grain production bases and the second largest cotton producer in Gansu Province (Niu, 2005). Agriculture and pastoral economy contribute to a large portion of the local revenue. As a result of the adjustment of the agricultural structures, the county's economy has continued to undergo a robust development of its cotton and sheep production in 2004. More than 200 square kilometres of land were devoted to cotton growing and the total output of lint cotton hit 32,500 tons. About one million meat sheep were raised to cater to the increasing demand of the national market. Farmers' annual per capita income stood at U.S.$520 in 2004, nearly half of which came from cotton production.

The plain areas facilitate the TV penetration and most of the villages in Jinta can be easily reached by TV signals. The number of TV channels Jinta farmers can assess range from a dozen to over twenty, including CCTV channels on news, entertainment and agriculture as well as the cable TV managed by the county TV station (at a monthly flat rate of RMB 10). TV serves as a major means for the local farmers to get informed and entertained.

The better off Jinta farmers have wider access to the print media as compared to their counterparts in other parts of western China. The majority of the interviewed farmers reported that they often read local newspapers such as *Gansu Farmer*, *Jiuquan Daily* and *Gansu Science and Technology*. The popular magazines mentioned include *Rural Youth* and *Chinese Women*. Newspapers are often shared in the rural areas. Despite the low per capita newspaper subscription rate, the interviewed farmers showed a stronger dependence on the print media for information about farming. This may be partly due to the relatively higher educational level of the farmers, most of whom are junior middle school graduates.

Another important channel by which Jinta farmers get exposed to agricultural information is the training courses and agricultural technology workshops organized by the county or the town government. These training courses are often held four to five times a year, especially at the high farming seasons, so that the agricultural experts and technicians can offer timely help when farmers experience problems in farming and breeding. In the sowing period, the experts often advise the farmers on new seeds and prices on fertilizers and machineries. When the sowing is over, the training usually switches to field management and pest prevention. Such training courses are very popular among the rural people.

9.2 The Change Agency

The county government started to set up websites in 1999. However, no effort was made to meet the information needs of the 110,000 farmers in Jinta, who form the overwhelming majority of the county population, until Shen Juntao became the county head in 2001. The transitions in the rural economy system and the changing role of the local government in terms of agricultural production activities serve as catalysts for Shen's initiatives of networking the rural households.

China's accession to WTO poses great challenges to the self-subsistence agriculture of the county, which has been managed as part of a planned economy for decades. The opening up of the rural economy makes it more subject to the influences of the market as the supply and demand of farm produces tend to fluctuate with market changes both at the national and the global level.

The inland geographic locations and the less developed economy put Jinta farmers in a disadvantaged position in adapting themselves to the market mechanism and gaining an edge over their competitors. The local economy has undergone setbacks resulting from the lack of updated market information. It is not uncommon that farmers' incomes decrease even with the land output going higher. As Jinta is not within easy reach of any of the national highways, its farmers constantly face the problem of finding potential buyers even when they have high quality farm produces. Due to limited channels of communication, it is difficult for the farmers to either adjust their farming plans in accordance with the market trends or establish business contacts with the outside. Therefore, keeping farmers informed about agriculture technologies as well as changes in the prices of farm produces is critical for them to get accustomed to the rules and practices of the market economy, which in turn may contribute to the transformation of the county economic structures.

When Jinta farmers are trying their best to cope with the changes brought about by the market economy, the local government has to shift its role from issuing administrative orders to providing service and consultancy to help farmers make informed production decisions. The new county head, Shen, realizes that the old practice of bureaucratic interventions can no longer work under the market mechanism and farmers may suffer huge losses due to production plans prescribed by the government. As the level of risk and uncertainty involved in the agricultural productions and management tends to be much higher under the impact of the market forces, the decision-making should be left with the individual farmers and let them be responsible for the possible losses or gains.

When studying rural informatisation, Shen noticed that with networked computers, the government agencies at the municipal or county level were in command of large amounts of up-to-date agricultural information. However, farmers, who made real decisions on production and management, were left ill informed due to their poor economic status, low knowledge level, and the geographical constraints. Shen believed the county government could serve the farmers better by setting up a rural information network.

After conducting door-to-door surveys in four villages near the county seat, Shen and the staff at Jinta Information and Network Centre decided to carry out the ICT project called "Networking the Villages and Informing the Farming Households" in 2001. The goal was to achieve the last mile connectivity through providing farmers with the information over the Internet.

9.3 The Diffusion of the Internet

The chief government agency promoting Internet diffusion is Jinta Information and Network Centre, which was set up as a branch under the County Bureau of Statistics in 1999. After the successful implementation of the "Networking the Villages and Informing the Farming Households" project in four pilot villages, Shen felt that the county needed an agency dedicated to the rural informatisation. As a result, the centre was separated from the Bureau of Statistics and reported directly to the county government.

The institutional change has not only granted the centre more authority over the township and village administrations but also helped to break the institutional barriers. Instead of serving a single government section, the centre is responsible for providing information services to all government organizations, private enterprises as well as individual farmers.

During the course of implementing the project, great effort has been made to minimize the cost and lower the knowledge barriers that the Internet technology poses to its rural users. Instead of making huge investments in purchasing servers and other network facilities, the Information and Network Centre adopts the concept of virtual servers and sets up its websites through renting the server at Lanzhou Telecommunications Bureau. The virtual server helps the centre avoid the possible expenses in technical maintenance of real servers and becomes a cost-effective solution to the extension of the Internet technology in this less developed county. With U.S.$120 as the annual rental fee, the 13 government websites run by the centre can enjoy the same network quality as users in big cities.

Instead of connecting every rural household, the Information and Network Centre uses an innovative way to let agricultural information reach the local farmers. The centre launches a biweekly E-newspaper titled *Jinta Jingji Xinxi Daobao (the Jinta Economic Development Herald)*. The content of the paper is compiled by two information technicians with one in charge of information collection and editing and the other responsible for typesetting and finalizing the paper into online versions. As locally born college graduates, both editors are quite familiar with the agricultural production and farmers' information requests for different farming seasons.

Much of the content in the E-paper comes from the national or local agricultural websites. It focuses on information about farming skills, fertilizers, seeds as well as the market supply and demand. According to its chief editor Li Fagang, the frequently visited websites include *Zhongguo Xingnong Network* (http://www.cnan.gov.cn/), managed by China Meteorological Bureau, which has branch websites in over 34 provinces and municipalities in China. Its branch in Gansu Province, *Gansu Xingnong Network* (http://www.gsnw.net/), supplies the *Herald* with large amounts of information on farming technologies, pest and disease control, fertilizers and pesticide, and current prices on farm produces, which bear close relevance to the local agriculture. Information on market analyses and trends can be obtained from *Beijing Agricultural Information* (http://www.agri.ac.cn/) and *Oriental Agricultural Information* (http://www.dfagri.com/, based in Zhejiang Province).

Government circulars and documents targeted at farmers serve as another source of information for the *Herald*. Before the existence of the E-paper, the information on farming and government policies from the local agricultural organizations such as the Animal and Husbandry Bureau and the Forestry Bureau had to go through various bureaucratic channels to reach the farmers. This often resulted in delayed transmission and also possible loss of information. The *Herald* now provides the government with an alternative way of spreading its messages among the rural people.

The feedbacks from readers are also taken into serious consideration when the editors are making preparations for a new issue of the *Herald*. While most of the responses come from phone calls, some are sent by the village information centres via e-mails. When an issue of the *Herald* is ready for publication, the online version will be uploaded onto the county government website (www.jtxnet.com). By clicking the hyperlink, one can either read the newspaper online or download it.

However, the objective of delivering updated online information to the farmers cannot be achieved simply by putting an E-paper onto the government website. Unlike the traditional way of agricultural extension channelled through government administrations, Shen and his staffers succeeded in persuading the village schools to play an active role in disseminating the information.

There are several advantages for the project to situate its information centres at the village schools. First, by circumventing the administrative channels such as the township government and the village committees, the Information and Network Centre can greatly speed up the time it takes for the *Herald* to reach the farmers. The majority of the local leaders are still clinging to the old thinking about governance[22] and therefore do not see the need of bringing information to the rural households. This would greatly hinder the effective diffusion of online information among the farmers.

Second, as the computer literacy is high among school teachers, they take the responsibility of downloading the online newspaper. This not only eases the shortage of qualified technical personnel but also helps to save the outsourcing and hiring cost. Third, many of the local elementary schools have existing networked computers as part of the nationwide modern distance education programme for rural areas sponsored by the Ministry of Education. By employing the hardware (networked computers) and the brainpower (teachers) at the village schools, the county government can easily promote the extension of online information at the village level without bringing extra financial burdens to the local farmers.

With the equipment of a printer and a mimeograph machine, the teachers can download the *Herald* from the government website and print it in hard copies according to the number of the village households. Then students can hand deliver the papers to their parents and neighbours.

In addition to its role of distributing the E-paper, the village information centre is also responsible for soliciting feedbacks from the rural readers. It keeps a log book recording the dates and number of newspapers distributed as well as farmers' responses. If a villager shows interests in a particular piece of information, he may call the Information and Network Centre directly or let a student bring his request to the teachers in charge of the operation, who in turn email the feedbacks to the county centre (see Figure 8). The village information

22 Despite the transition to market economy, the rural leaders still feel reluctant to shift their roles from issuing administrative orders to providing service and consultancy to help farmers make informed production decisions.

centre also helps the Information and Network Centre conduct surveys to monitor the information requests of the farmers and evaluate the relevance of the information in the *Herald* to the local agricultural production.

For those remote schools excluded from the distance education programme, the county government and the local village committees often jointly cover the cost of a computer, a mimeograph machine and a printer. These village schools often access the Internet through a phone-dialling mode. Except the investment in computer hardware, the cost sharing operation allows the county government to pay less than U.S.$100 a year for each village to receive the relevant information from the Internet (mainly for the cost of paper and ink). With the initial success at four pilot villages in 2001, the operation soon spread to 75 villages, covering 90% of the rural households within four years.

Our field observations suggest that the printed copies of the *Herald* have resulted in vast diffusion of the online information on agriculture and the rural market. The *Herald* has become one of the main sources of information for local farmers to make production decisions.

9.4 The *Herald*

To better understand the reasons behind the rapid diffusion of the E-paper, a content analysis was conducted by the researcher on the accessible issues at Jinta government website, with 114 issues of the *Herald* published between April 2001 and October 2005.

The results of the analysis suggest that the *Herald* has been finely attuned to the local information requests as conditioned by seasonal and geographic characteristics. Altogether 1,254 information items (as measured by the headlines) are covered within almost five years of circulation, with each issue carrying 11 information items on average. The most popular categories include farming technologies (744 items, 59.3%), market analysis (271 items, 21.6%), supply and demand posted online (150 items, 12%), public health (59 items, 4.7%), government policies (18 items, 1.4%), and weather forecast (12 items, 1%).

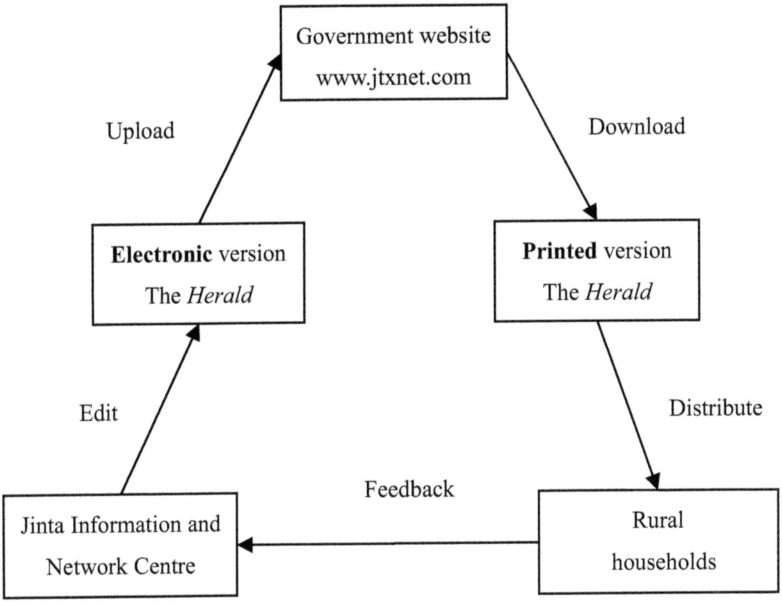

Figure 8. The diffusion of the web-based information at Jinta County.

With respect to farming technologies, the skills and tips on livestock raising, disease control, greenhouse cultivation, and fertilizers and pesticides carry much more weight than those on crops, cotton, fruit cultivation and new breeds (see Figure 9). The variation in coverage reflects the transitions taking place in the county economic structure. As the crop yields have almost reached the maximum after hundreds of years of cultivation, farmers start to turn to sidelines such as livestock rearing (sheep, cows, pigs or chickens) or greenhouse vegetable growing for more income.

The 21st, 22nd and 23rd issues of the *Herald* in 2003 are special editions related to the raising of sheep, pigs and cows. The sheep edition told farmers the problems they should be aware of if they wanted to raise sheep on a large scale. Farmers were also taught about how to tell the age of a new breed of sheep and help the lambs put on weight after the breast feed is over. The pig edition discussed eight kinds of diseases easily contracted by pigs and provided details about the symptoms and ways of treatment. The cow edition was mainly on the disease control by using medical herbs.

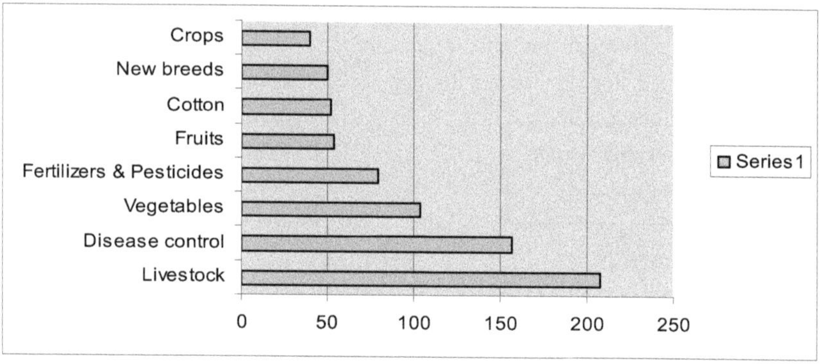

Figure 9. The allocation of subcategories under the farming technologies.

With the overall improvement of people's living standards, there has been a great surge in the demand for fresh vegetables in winter. More and more Jinta farmers are engaged in producing off-season vegetables. Instead of idling themselves and their land for almost three to four months in the winter season, the northern farmers are able to break the climate constraints and work all year round, provided that they are equipped with the modern farming techniques. The adoption of new ways of farming involves skills acquisition so as to reduce the level of risks and uncertainties associated with the investment ventures. Therefore, the *Herald* devotes much space to the advanced greenhouse cultivation technologies, such as biological pest control as well as the temperature and humidity management.

Moreover, the *Herald* offers farmers much advice about the disease control for plants as well as animals. Because of a shortage of agricultural experts in the countryside, farmers often suffer great losses if the diseased cotton plants or animals do not receive timely treatment. Based on the agricultural websites containing detailed descriptions of the recently discovered diseases and the most effective treatment, the *Herald* helps farmers make accurate diagnosis and take appropriate steps to manage the problems.

In addition to exposing farmers to the updated agricultural technologies, the *Herald* also keeps farmers informed of the price fluctuations as well as the market trends. The results of the analysis suggest a high weightage has been assigned to reports on cotton, sheep and other crops. This slant of coverage coincides with the local development needs as cotton and meat sheep production serve as two stanchion revenue-generating sectors for the local economy. One third of market analyses are about cotton prices or the supply and demand in either the domestic market or the international market (see Figure 10). The market information functions as indispensable references for farmers to make wise farming decisions and adapt to the market changes.

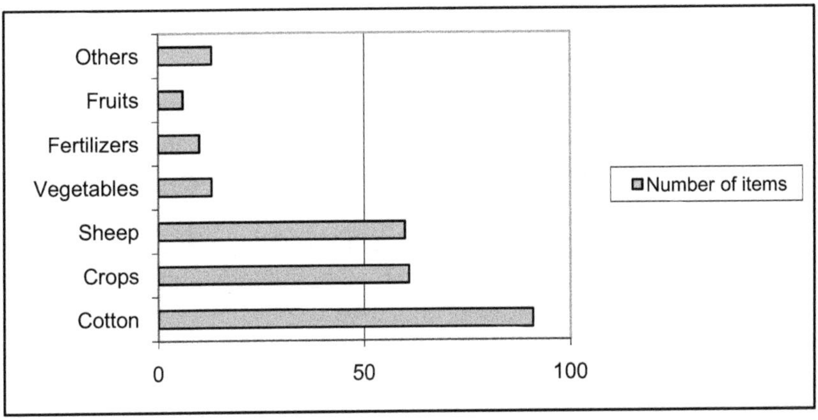

Figure 10. The allocation of subcategories under the market analysis.

In 2001, the *Herald* reported that the cumin harvest in India and Pakistan suffered a great loss due to a natural disaster. The county farmer association quickly collected cumin at a higher price and sold directly to an Indian company dealing in agricultural products. This helped the local farmers make an extra earning of RMB 8 million totally.

The *Herald* acts not only as an information provider and advisor but also as a communicator through establishing a business platform where local farmers and enterprises can exchange product information and establish contacts with the potential business partners. The information about the agricultural products is categorized into supply and demand when it is posted online. Of 150 posted items, 81 are for sales, 60 for purchases and nine seeking local partners. Generally speaking, the advertised products include cotton, crops, vegetables, livestock and medical herbs.

Jinta Information and Network Centre does not charge local farmers or external parties for publicizing their agricultural products in the *Herald*. Besides a detailed description of the products, the *Herald* also provides contact information of farmers or traders. For example, many villagers of Sanwangou Village were interested in a piece of information about a special breed of Han sheep in far away Shandong Province. Many previous reports in the *Herald* advised the local farmers to adopt the breed because the weather condition in Gansu is quite suitable for raising the sheep. However, the farmers remained sceptical about the information. Only after they saw the pictures of the sheep farm over the Internet did they get convinced. Using the contact information printed in the *Herald*, the village managed to buy 180 sheep from Shandong, which turned out to be their cash cows.

Though the *Herald* devotes much of its content to information related to agricultural production, it is also responsible for propagating government policies and health information to the rural people. The *Herald* often runs columns on national and local policies closely related to the rural life, such as taxation reform, subsidies for grain planting, tax relief for natural disasters as well as welfare packages for households with no male offspring. Prior to the existence of the *Herald,* TV has been the only medium for such a purpose. However, during the high farming seasons, villagers are too busy to watch TV, which is not particularly good with details anyway. To meet the farmers' needs, the *Herald* often carries speeches by government officials and government documents of relevance to the farmers in their entirety, thus opening an additional channel of information on rural policies and improving the transparency of the government.

During the outbreak of SARS in 2003, the April and May issues of the *Herald* carried large amounts of information about the prevention of the epidemic. The farmers learnt of not only the various transmission modes of the disease and its typical symptoms but also the necessary precautionary measures. The paper also warned the local people not to take the traditional herbs without consulting the doctors. The 11th issue of the *Herald* in the same year was a special edition aimed at creating awareness of family planning. The reports ranged from the appropriate uses of contraceptives and government birth control policies to the newly founded family planning association. Despite its infrequent coverage, the health information can supplement the existing channels of communication and help to disseminate public health messages and disease prevention techniques as well as promote healthy lifestyles among the farmers.

In order to make the *Herald* better serve the development needs of the local agriculture, Jinta Information and Network Centre tries its best to make sure that the coverage in the *Herald* matches the farming calendar and the cycle of crop growth under the climate of northwestern China. Despite being a so-called bi-weekly publication, the issuing of the E-paper does not actually follow the timeline rigidly and enjoys much flexibility to cater to the farming cycles. Jinta farmers receive the *Herald* more frequently at the sowing and harvest time, whereas the non-farming seasons show an apparent fall in terms of the number of issues. Therefore, the publication of the E-paper is seasonally based rather than monthly based (see Figure 11).

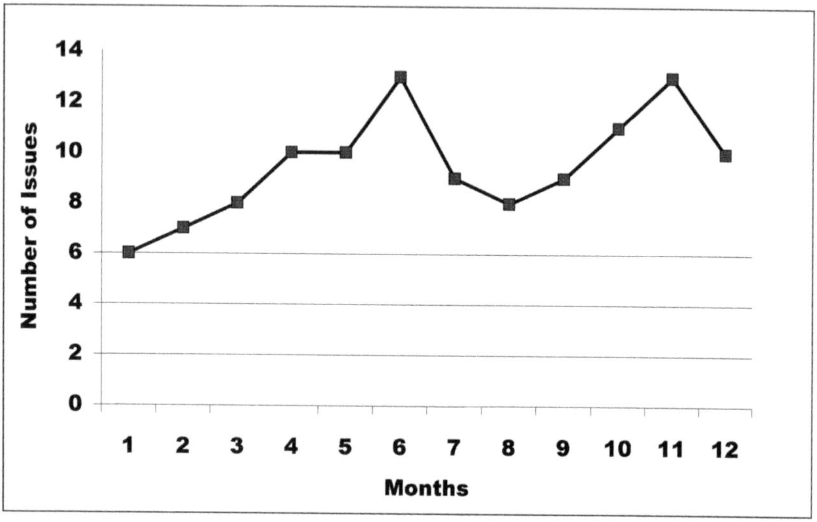

Figure 11. The issuing of the *Herald* on a seasonal basis.

Much of the information in the *Herald* is selected to synchronize with the changing farming seasons. This has been supported by the analysis of the spread of various information categories on the seasonal basis (see Table 6). During spring sowing, farmers read most reports on new breeds of crops and livestock, fertilizers and the least market information. In summer farmers are more often informed of the ways of disease control for their livestock. In autumn when the produces are ready to go to the market, the *Herald* often carries most information on the prices of various farm products at major markets. In winter farmers are taught about the preservation of grains and vegetables and the operation of greenhouses.

Table 6
The Analysis of the Newspaper Content on the Seasonal Basis[23]

Categories	Spring	Summer	Autumn	Winter
Market information	Least		Most	
New breeds	Most	Least		
Disease control	Least	Most		
Fertilizers & pesticides	Most			Least
Vegetables	Least			Most

9.5 The Experiences of Jinta and their Implications

Jinta County has witnessed much social and economic progress with the implementation of the "Networking the Villages and Informing the Farming Households" project. There has been a consistent growth in the annual per capita income for the local farmers, rising from U.S.$435 in 2001 to U.S.$520 in 2004. On average, Jinta farmers' personal income is 2.3 times that of the farmers in the province on average. According to the county statistics, 12 towns in Jinta County secured business transactions of over RMB 1 billion by posting sales information on the Herald as well as the local government websites in 2003 (Shen, 2004).

The success of Jinta enables the county to take the lead in diffusing the Internet technology in the province. Gansu Development and Reform Commission has decided to replicate Jinta's practices by set-

23 According to the local calendar, spring is from March to May. Summer is from June to August. Autumn is from September to November. Winter is from December to February.

ting up a rural information service network covering the whole province. The provincial government will invest RMB 88.5 million to build 87 county information centres and 15,000 village school information centres in the next five years (2005–2009), with each county having approximately 60 village information centres (GDRC, 2005). The State Council Informatisation Office has taken Jinta's experience as one of the most cost effective models of innovatively employing the Internet to provide farmers with information services and started to investigate the possibility of extending the practice to other rural areas with relatively low level of development (Shen, 2004).

The positive impact of the *Herald* on the farmers' life can be easily observed. The advanced farming skills and the updated market information help farmers raise crop output and livestock yields, which bring them better income. A farmer named Wang Qing in Jinda Village reported that the warning of severe weather changes in the *Herald* helped him avoid damages caused by windstorms and frost to greenhouse vegetables. Zhu Xuming, a farmer in Hongguang Village, secured a high output of cotton by planting a quality breed recommended by the *Herald.* As the new breed was more suitable for the local climate and geographical conditions, Zhu ended up with producing a bumper harvest way ahead of other cotton producers.

Many factors contribute to the rapid diffusion of the Internet technology and the widespread adoption of the online information among Jinta farmers. First, the Jinta project is mainly a result of the government initiative. The diffusion of the Internet was included by Shen in the development agenda of the county government from the very beginning, which led to the subsequent planned and organized extension efforts both at the county and the village levels. The Jinta Information and Network Centre, directly affiliated to the county government office, became the chief agency responsible for not only carrying out the project at the county level but also supervising, coordinating and evaluating the implementation work at the village level. On the one hand, the information centre serves as a technical institution by selecting and compiling online information and uploading the *Her-*

ald onto the Internet; on the other hand, it is a government agency with administrative power over the town and village administrations.

When the local economic development and the rural telecommunications infrastructures are not ready to sustain the natural diffusion of an advanced technology like the Internet, only a powerful institution like the government can mobilize social resources to push the technology to its rural users. The multi-million Yuan investment in installing computer and network facilities and covering the maintenance fees and the printing costs at the county and village information centres could only come as a result of the commitment of the government. If it were not for Shen as the county head, the village schools would not agree to play a part in the project, which apparently adds extra workload to the school management.

Secondly, the change agency has made wise use of the Internet technology to diffuse online information from the county to the villages. Instead of going through the traditional agricultural extension channels such as the township government and the village committees, the transmission of the *Herald* can be completed within seconds by uploading and downloading information over the Internet. The Jinta project has successfully bypassed the institutional hierarchies and greatly facilitated the diffusion process.

Thirdly, the project has been carefully designed to minimize the costs and ensure its sustainability. The adoption of the virtual server at the county information centre greatly reduces the hardware expenses and makes up for the technical inadequacies. The pre-existing computer facilities and teachers with IT skills in the village schools also help the county government avoid the possible costs involved in the purchase of computer and network facilities as well as the employment of technical personnel. The students, on the other hand, serve as the best messengers due to their natural links to the farming households. The last mile connectivity is innovatively achieved without building telecommunication infrastructures catering to a limited population dispersed across large areas. With an annual expenditure of RMB 60,000, the county government manages to extend the updated agricultural and market information to about 90% of the households in the county.

Fourthly, several characteristics of the *Herald* contribute to its wide adoption and popularity among Jinta farmers. Rogers (1983) argued that an innovation tends to diffuse rapidly within a social system when it is of relative advantage, compatibility, less complexity, triability and observability. The information needs of Jinta farmers are greatly enhanced as the county economy is getting more market-oriented. The adoption of scientific ways of farming and new breeds of crops and livestock requires advanced farming skills as well as changes in traditional farming. However, the existing communication channels such as TV cannot meet such information requests due to the shortage of programming content targeted at the rural people. The *Herald* has become an indispensable source of information for the local farmers to make decisions on agricultural production and management.

The *Herald* also manages to target its content at the local farming cycles. Not only the number of issues but also the content of the newspaper has been finely attuned to the varying information needs. Moreover, the transformation of the *Herald* from a digital medium into a print medium circumvents the requirement of IT knowledge, allowing Jinta farmers to access online information while staying offline.

Moreover, the *Herald* integrates the advantages of both the old media and the new media. The employment of the digital technology enables the fast collection and transmission of the agriculture information over the Internet. Its conversion into the print format greatly reduces the cost and the knowledge barriers posed by the advanced Internet technology to its rural users. The printed words give farmers more leeway in terms of the time and venue to read the information. This has greatly enhanced farmers' exposure to the information on the Internet. Its free access, easy readability and portability lead to the wide adoption of the E-paper among the rural people.

The objectivity of the information source also helps to explain the sustained adoption of the *Herald* among Jinta farmers. According to the deputy Party secretary of Zhongdong Town, nowadays farmers no longer believe in the information from the government circulars. In the past, the local government often interfered in the cultivation decisions of farmers by issuing administrative orders. Consequently, the farmers' interests were sacrificed to meet the goals set by the local government. For example, in order to prevent a sugar factory from going bankrupt, some farmers were forced to plant sugar beets despite the low profit. Therefore, farmers are inclined to follow the *Herald* rather than government directives.

Without direct access to the Internet, the printed copies of the electronic paper enabled 90% of the local rural households to obtain updated agricultural information from the Internet at relatively low cost for four years. Jinta's experiences indicate that as the Internet usage involves a fair amount of technical know-how and the purchase of expensive computer hardware, online information is unlikely to flow to the farmers in one step. The additional steps in the flow of web content to Jinta farmers actually help to reduce the costs of Internet access, overcome the technical barriers, and facilitate the processing of information.

The findings of the study reveal that the two-step flow model, a concept developed through the examination of the role of mass media in political communication, may apply to the process of transmission of online information in rural areas where technical and knowledge gaps need to be bridged. The original two-step flow model holds that the communication messages first flow over mass media channels to opinion leaders, who in turn pass them on to the less active information seekers. Likewise, online information may go through multiple nodes before it is disseminated to the rural people. The two-step or rather the multiple-step flow of online information is an effective way to help farmers cross the cost and knowledge barriers posed by the Internet.

The innovative practices of Jinta have had important implications for narrowing the digital divide between the urban and rural areas. With the overall economic and infrastructure contexts unfavourable to the wide diffusion and adoption of the Internet in the rural areas, the two-step or multiple-step flow model can allow rural people to access updated online information on market trends, advanced farming technologies, education and healthcare without going online. This will contribute significantly to reducing the widening socioeconomic disparities between the information haves and information havenots.

10. Tongnan County

10.1 The Profile

The Project on Poverty-Reduction through Access to ICTs, jointly sponsored by the Chinese Ministry of Science and Technology (MOST) and UNDP, hoped to bridge the digital divide between the urban and rural areas by setting up information and communication centres in five poor counties with varying geographic, social and economic conditions. This chapter focused on Tongnan County of Chongqing Municipality, which was said to have achieved the most impressive results among the five counties according to the UNDP mid-term evaluation report (Ulrich, 2003).

Despite its size, Tongnan has the least developed economy among the five counties[24] covered by the project, with a population of 903,955 residing in its 31 towns and 281 villages. Tongnan lies in the northwest of Chongqing and to the middle east of the Sichuan Basin. The rural residents account for 88% of the county's population. Whereas the annual per capita GDP stood at around RMB 7,380 (U.S.$900) at the county level, the equivalent figure for the farmers was RMB 2,805 (U.S.$342) in 2004. This partially reflects the sharp social and economic disparities between the urban and rural areas within the county. According to the latest figures released by the County Statistics Bureau, almost 70% of rural people's income went

24 The five counties covered by the pilot project are Yuyang district of Shanxi Province, Wu'an city of Hebei Province, Huoshan County of Anhui Province, Shangcheng County of Henan Province and Tongnan County of Chongqing Municipality (Ulrich, 2003).

to living expenditures and the Engle's coefficient remained 58.3, with the rest 30% invested in production.

The county's terrain is dominated by hills and flatlands, with Fu River and Qiong River running through the district and a drainage area of over 100 square kilometres. The unbalanced access to natural resources has resulted in different levels of economic development. The towns and villages bordering the rivers and streams benefit from the fertile land with abundant water for irrigation, while the hilly areas have to rely on rain water to sustain farming and the local residents can barely make ends meet due to the low yields.

The Tongnan County office for the UNDP project selected two impoverished towns, Shanghe and Red Flower, for the experiment and each town set up two village Internet centres to cater to the information needs of local villagers. The background information of the two selected towns is provided below to introduce the social context for this Internet diffusion project.

Shanghe is situated in the east of Tongnan County, with 10 villages and about 30,000 people under its direct administration. Its annual per capita income was RMB 2,050 (U.S.$250) in 2004. Red Flower, which lies in the north of the county, administers 6 villages with about 16,000 residents. The 2004 annual per capita income of the local farmers stood at RMB 2,014 (U.S.$246) (see Table 7).

The small-scale farming economy is prevalent at the project sites, with self-sustained cultivation and livestock rearing as the chief modes of production. The major land produces include rice, wheat, sweet potatoes, peanuts and rapeseeds. Almost every family keeps a few pigs, which bring in about 40% of the household income. Half of the labour force in Shanghe have left their land and migrated to urban areas for job opportunities and the corresponding figure for Red Flower is one third. In 2004, the income generated by these migrant workers accounted for 29% of the local revenues in Shanghe and 24% for Red Flower.

Table 7

Basic Information about Shanghe Town and Red Flower Town (2004 Statistics)

	Shanghe Town		Red Flower Town
Number of villages	10		6
Total population	30,000		16,000
Rural population	27,330 (91%)		15,347 (96%)
Labour force	Female	7,083	4,050
	Surplus	103	2,130
	Migrant workers	8,053	3,210
Per capita income (RMB)	2,050		2,014
Total revenue of rural economy (RMB)	Household	56 million	35 million
	Collective	0	0
	Migrant workers	24 million	13 million
Dropout rates of elementary school	3%		3%
Dropout rates of junior school	2%		5.6%

In Shanghe Town, the fixed-line household telephone penetration (10%) is much lower than that of Red Flower Town (31%), but their TV ownerships are about the same (a little bit over 70%). The majority of the rural residents in both towns have an average of six to seven years of schooling. Thanks to the improvement of the local transportation, both towns can reach the county seat by an hour's ride. However, the village roads are poorly maintained and can get quite muddy on rainy days.

10.2 The MOST-UNDP Project

The MOST-UNDP project followed a typical top-down approach, which involved a four-tiered support structure (see Figure 12). In this U.S.$2.5 million poverty alleviation initiative, one quarter of the funding was provided by UNDP for training and consultancies, with MOST and the five benefiting counties sharing the rest of the costs. The county government in Tongnan contributed U.S.$200,000 to the total investment. The project officially commenced in March 2001 and terminated at the end of 2003 when the funding was used up.

The Rural Agriculture Technology Development Centre of MOST and UNDP were jointly in charge of the operation and supervision of the project, whereas the Ministry of Commerce (MOC) acted as the financial controller. All the hardware facilities were purchased from Beijing and sent to the individual counties through airfreight. The Beijing headquarters provided trainings to technicians from the project sites and sent experts to guide and evaluate the implementation work at these participating counties.

The primary motivation for the county government to apply for the Internet project was to get the funding. It was hoped that the government office facilities, both at the county and township levels, could be upgraded and websites could be set up to enhance government efficiency and facilitate the daily administration. The county UNDP office was equipped with networked computers, photocopiers, printers, digital cameras as well as projectors to organize computer training for information technicians from the surrounding towns and villages (Z. Tang, personal communication, May 16, 2005).

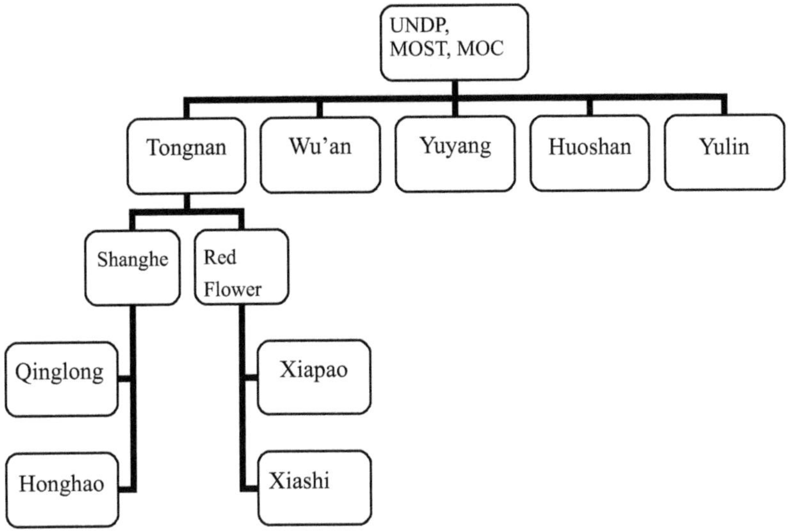

Figure 12. The structure of the ICT project.

During the first year of the project, most efforts were spent in recruiting information technicians, conducting trainings and creating project awareness among the local farmers. As there was a severe shortage of computer literates, the county UNDP office invited experts from outside to teach basic computer skills. Despite the fact that the majority of the 24 information technicians had no prior knowledge about computers and the Internet, they managed to learn how to type, search and post information online, and handle emails within a month of training. An advertising campaign was launched to promote the project, using the local newspapers as well as radio and TV broadcast. Wall posters, bus banners and sign boards were also put up to target the rural residents not accessible by the print and electronic media.

The UNDP centres at the townships and villages started operation in late 2001 with two computers, a printer, a photocopier and a fax machine installed in each of the project sites. The town centres were initially connected to the Internet via a modem, which was later replaced with the Asymmetric Digital Subscriber Line (ADSL) service provided by China Telecom in late 2003 (H.Li, personal communication, May 17, 2005). As the Internet centres at the township and village levels were built with the same design, only the centres at Shanghe were described here in detail.

The UNDP project centre of Shanghe was located inside the township government building. It was supposed to cater to the information needs of farmers from all the surrounding villages. Affiliated to the local government, the two information technicians worked only part-time at the centre, which opened half a day when the villagers gathered for the town fair held in every three days. The township UNDP centre served as a link between the village centres and the county office. The information technicians at the township centre offered technical support whenever the computers at the village centres were down.

All the four village Internet centres were located inside the primary school at each village. Each centre had a dial-up line for accessing the Internet from one of the two computers. The mud-bricked house, wooden benches and barefoot students offered a sharp contrast to the wired computers sheltered under the same roof. The village centres played an important role in the Internet diffusion and usage in rural areas as they provided farmers with the opportunity to directly operate the computers and surf the net.

In Qinglong Village of Shanghe Town, the village head and a teacher of the village primary school served as the information technicians for the village centre, whereas in Xiashi Village of Red Flower Town, the village head and a better educated farmer acted in that capacity. Outside the village centres there were posters to publicize the UNDP project as well as useful agricultural information gleaned from the Internet. The village centres were committed to conducting information searches for farmers who could not type and helping farmers and students acquire the computer and Internet application skills.

The trainings were carried out in several phases to target different groups of people. Young farmers with some years of schooling or outside working experiences received computer training first. Then it was the turn for young women who were followed by the middle-aged or elder farmers with an interest in the Internet. The training was finally given to students in the primary schools.

Though the young schoolers were the last to receive the training, they were the quickest to adopt the technology. It took longer for the older generation to get rid of their technophobia. Many adults were unwilling to scrub the mouse against the pad, believing that constant friction might make the mouse less durable. The quick adoption of the Internet among the kids resulted in the phenomenon of "proxy usage" (Ulrich, 2004, p.20), as much information searching was conducted by the school kids for their parents.

The diffusion of the Internet was combined with the introduction of other ICTs to the local villagers. The fax machines, videotape recorders and VCD players served as supplementary tools to aid the dissemination of advanced farming skills and facilitate rural people's communication with the outside.

A prominent feature of Internet diffusion in Tongnan is that the promotion of the Internet technology and E-literacy went hand in hand with the extension of agriculture knowledge by utilizing the existing channels of communication. In view of the overall low level of education among the local farmers, most of the efforts were not spent in training the farmers to acquire E-literacy but in creating awareness about the Internet and increasing farmers' access to advanced farming skills. The ICT project was taken as an opportunity to spread agricultural knowledge through the traditional modes of agricultural extension.

Be it print or electronic, every channel of communication was employed by the project to raise awareness about science and technology among the rural people. The scientific films were produced and TV programmes on agricultural knowledge were recorded. Handouts were compiled using information from the existing publications targeted at farmers, such as rural people's handbooks and questions and answers about law and technology. The Internet centres at the township and the village levels organized the villagers to watch films or VCDs together or allowed them to borrow for later viewing at home. The handouts were often distributed at the collective viewing sessions.

The agricultural extension efforts also took the form of interpersonal communication. Within the project period, MOST sent 17 scholars and experts to offer lectures to the local farmers, on topics ranging from the implications of China's accession to WTO on rural economy to the latest findings in agricultural research. The county UNDP office organized 150 fairs to propagate information about the project and help farmers seek advice from agricultural experts.

10.3 The Internet Use

By employing the server provided by the MOST-UNDP project, the county centre set up a science and technology website (www.xtkj.net) in May 2002 to publicize information about the project and promote the applications of the Internet technology. Both the town and village centres could help farmers apply for free email accounts and post supply and demand information on the county website, which greatly facilitated the information sharing among the UNDP centres at the county, town and village levels and functioned as a window for the county to be known to the outside world. By the end of 2003, there had been 66,000 website visits, of which one third were made by users from towns and villages.

At the town and village Internet centres, the Internet was employed as an efficient tool in searching information for farmers and helping them advertise their farm products online. During the non-farming seasons, the centres were crowded with farmers, whose information requests ranged from prices of farm produces to cultivation skills, and from health care to external job opportunities. Most often the online searching results would be printed free-of-charge and the technicians would also help to explain the findings to the farmers. Handouts of farming information downloaded from the net were also distributed to the farmers. The most frequently visited websites were the local agricultural and government websites. The farmers found the information service very useful and practical. For example, they would come to the centre for information when their pigs contracted an unknown disease and the veterinary was not around, or for information regarding the sanitization of pens and the right additives to the forage.

In order to find out the online information that local villagers need to guide their farming activities, the researcher conducted a content analysis of the information searches from November 2002 to October 2003 at Qinglong Village centre. A total of 206 items in the Internet search log kept at the period was included in the analysis. The results of the analysis show that the rural users perceived the Internet as a positive tool with the potential to help them enhance agricultural productivity. Almost 90% of the information searching was about farming activities (see Table 8). Information on farming skills, market prices, fertilizers, pesticides, disease control, and new breeds was among farmers' top concerns. Cash crops such as rape seeds, peanuts and soy beans dominated the inquiries about the crops. As far as information on livestock was concerned, most of the requests centred on poultry or pig rearing techniques, disease control and prices at the local markets.

However, as agricultural production differed from region to region, farmers often met difficulties in processing online information even though they enjoyed easy access. Therefore, the town and village centres acted as information gatekeepers by helping farmers seek the most useful information over the Internet and assuring them of its relevance to the local farming. As the central links in the project structure, the Internet centres played a significant role in raising awareness about the Internet technology among the farmers and showing them how to use the Internet to their advantages.

Table 8

Information Search at the UNDP Centre of Qinglong Village (November 2002 to October 2003)

Information categories (206 items)		Number of items
Farming skills (85 items, 41.3%)	Crops	28
	Vegetables	8
	Livestock	49
Market information (50 items, 24.3%)	Crops	25
	Vegetables	2
	Livestock	23
Fertilizers, pesticides & disease control (31 items, 15%)	Crops	18
	Vegetables	1
	Livestock	12
New breeds (18 items, 8.7%)	Crops	17
	Vegetables	0
	Livestock	1
Travel & Transport (12 items, 5.8%)		12
Health (2 items, 1%)		2
Job hunting (3 items, 1.5%)		3
Education (3 items, 1.5%)		3
Others (2 items, 1%)		2

The Shijing Village of Shanghe Town had a large piece of fertile land. Owing to the local farming traditions, the villagers used to grow dry land crops with low yields. A young farmer named Wu Xudong believed that he would generate more income if vegetables were planted. Therefore, he asked for advice from the town information centre. The technician searched the relevant information over the Internet and contacted the County Vegetable Office for assistance. The Vegetable Office recommended to Wu a new breed of green onion, which was suitable for growing under the local weather conditions. Benefiting from the advice offered by the county experts as well as the online information about vegetable management, Wu managed to earn over RMB10, 000 per year after renting the land and growing green onions on it (H. Li, personal communication, June 10, 2005).

Two farmers of Shanghe Town were interested in setting up a joint venture in raising rabbits. They went to the town Internet centre for help since they did not have any skill or experience whatsoever. After searching information over the agricultural websites, the technicians found out that the local market demand for rabbit meat was pretty low and the disease treatment for rabbits could be very complicated. As a result, they advised the farmers to raise sheep or chickens, which demanded less skill. In 2003, the two farmers made about RMB 30,000 by rearing more than 300 sheep (H. Li, personal communication, June 10, 2005).

A farmer named He Longhai of Red Flower Town raised a dozen cattle. He often went to the village Internet centre to get information on cattle rearing techniques. However, he had difficulties in applying the skills. Under the encouragement of the information technicians, he participated in a training programme on livestock rearing organized by the County Poverty-Reduction Office. The training enabled the young farmer to grasp many useful techniques. He changed from free range to pen rearing, which proved to be a cost effective way of raising cattle (M. Li, personal communication, June11, 2005).

The farmers' experiences showed that the gate-keeping functions of the town and village centres were actually accomplished with the backup of the agricultural institutions. Even though these institutions were supposed to serve the farmers' development needs with the expertise and capital at their command, no efficient channels of communication were established for farmers to directly employ these social resources to benefit their production. The information centres seemed to fill this gap by letting the agricultural institutions be aware of farmers' requests and ensuring that effective measures would be taken to address these requests. The information filtering process helped farmers come up with informed farming decisions and greatly reduced the uncertainties and risks involved in adopting the online information.

Despite the popularity of the Internet, the outreach of the technology only touched a small portion of the local farmers. According to the county UNDP office, around 20,000 people used the Internet services provided by the town and village centres within two years of operation, accounting for about 2.2% of the county population (Y. Xiang, personal communication, June 12, 2005).

Our findings suggest that the pre-existing gaps in social economic status, education and age resulted in a differentiated access to the information services among the rural users and consequently unequal benefits from the new technology. Even though the average income level of farmers at the project sites is less than one US dollar a day, treating them as a homogeneous group will easily lead to a skewed view of reality.

As the village Internet centres were under the charge of each village head, the village committee members and their friends and relatives tended to have more access to the information service. Young people with a strong interest in science and technology, business-oriented farmers, and school students were among the active users of the Internet. The fact that much of the information searching was accomplished by the schoolers on behalf of their parents lends further support to the postfigurative culture proposed by American anthropologist Margaret Mead (1970). The older generation was far lagging behind the young in terms of E-literacy and therefore had to rely on the young to harness the technology for development purposes.

In view of the large rural population at the project sites, farmers who benefited financially from the Internet services were by all means a tiny minority. Ulrich's (2004) social impact assessment survey conducted at the five UNDP counties showed that the estimated financial benefits received from the Internet centres were found to be positively correlated with farmers' confidence in the online information and their positive attitudes about the usefulness of the technology. Our findings at Tongnan supported the survey results. Those farmers who seemed to have benefited most from the information service in monetary terms tended to be business-oriented and think favourably of the Internet.

Liu Xiaolan ran a grain shop in Qinglong Village. She bought rice, peanuts and rapeseeds from fellow villagers and sold them at a higher price. Prior to the existence of the Internet centre, Liu depended on fellow villagers who frequented the county seat for information about grain prices. Often inaccurate messages resulted in big losses for her. Naturally, the Internet became a reliable source for her to know the prices for various farm produces. Liu visited the centre every other day to ask the technician to search the Internet for updated prices of farm produces as well as analysis of market fluctuations. When she learned from the Internet that the price of the rapeseeds was going to drop, she immediately sold her stock at a reduced price. She also warned other traders about this, but none of them believed in her and as a result they suffered a big loss when the price fell dramatically a few days later.

The better-off farmers who engaged in the production of marketable goods or ran an enterprise seemed to rely heavily on the market information and thereby generated more profits by using the Internet services. Getting informed of market prices as well as posting the farm produces online enabled the farmers to assess the real market value for their farm products and cut the chain of selling through direct contact with the buyers. In contrast, the Internet did not seem to bear much relevance to the life of rural people whose land produces were barely enough for their own consumption.

Xiashi Village was known for its knitted bamboo sheets, a handicraft passed on from one generation to another in the village. The product used to be sold within the province only due to the lack of information about the markets elsewhere. With the help of the Internet, the villagers could sell their bamboo sheets to other parts of the country by posting information about their products on the county website. The potential buyers could contact the farmers and place their orders by calling the phone number listed on the website. One of the villagers engaged in the business estimated that he earned U.S.$1,000 more a year as a result of this online help (H. Jiang, personal communication, May 18, 2005).

10.4 The Issue of Sustainability

The MOST-UNDP project came to an end in late 2003 after the funding was drained and activities at the town and village centres soon began to dwindle. The UNDP county office still existed as part of the county government, whereas the township and village Internet centres all closed their services in mid-2004.

Today the excitement about the Internet all died out except in the local farmers' account of their experiences with the technology. At the county seat and the two townships, all the computers and network facilities installed for the project are now used by the government offices. In Xiashi Village, the fax machine is used by the village head as a telephone and the computers are locked up in a warehouse. In Qinglong Village, the village head keeps the computers in his house, sitting idle. Some of the information technicians trained for the project have left their hometown and found jobs elsewhere.

Life of the farmers has gone back to the normal. Some farmers attribute the collapse of the Internet project to the termination of the funding, while others see themselves as guinea pigs for experimentation and consider it natural for the centres to close since the experiment is now over. For those farmers who have derived considerable financial benefits from the information services, the closedown of the Internet centres makes them feel handicapped because the updated information is no longer available. Nevertheless, they do not see anything that they can do to change the situation.

The short-lived ICT project prompts us to examine the social factors that may work against the sustained use of the Internet in the rural area under study. The findings of the study suggest that the rural land system and the modes of production are crucial factors determining farmers' information needs, which in turn shape the adoption and use of the Internet. With the adoption of the household contract responsibility system, the land allocated to each household has to be further shared by the male offspring and the tension between the shrinking arable land and the ever growing rural population becomes the root of many problems. In Tongnan's case, the per capita arable land, most of which is on high plateau, averages only 0.9 mu (about 600 m^2). The limited land plus the hilly landform does not allow mechanical farming as in the Western countries. As a result, manual labour aided by animal power and primitive farming tools is still the dominant mode of production. Farmers produce mainly for their own use and they need to engage in sidelines such as raising pigs to supplement the family income.

With such a primitive farming economy, farmers in Tongnan do not need a big market to sell their produces and therefore experience less needs for information. Even though most of the farmers are aware of the advantages offered by the Internet and the financial benefits it has brought to some of the early adopters; they cannot sense an immediate connection between the technology and their life. The small-scaled farming allows the majority of the local people to produce merely enough for the family to consume and exchange for other daily necessities at the town fairs. This explains why the Internet only appeals to the rural residents engaged in trading farm produces whose needs for market information are greater than their fellow villagers.

Though the limited information needs impede the local farmers' spontaneous efforts to continue the adoption of the Internet technology, the collective Internet access could still remain financially viable if the township government and the village administration would chip in. With the computers and other network facilities already in place, the monthly expenditure of a village information centre was only around U.S. $25 to cover the electricity, phone bills, Internet access as well as printing fees.

However, with the implementation of the household contract responsibility system, collectivism championed by the former communes has been replaced by individualism as farmers depend on their personal efforts rather than collective action for economic development. Moreover, with more and more farmers leaving their land to join the migrant labour force in big cities, the village administration no longer acts as a leader and planner for rural development and cannot interfere in the farming decisions of the villagers.

On top of that, the new rural taxation system taking effect in 2004 dealt another heavy blow to the funding of the village committees and town government. Under the new system, the agricultural tax should be reduced gradually until full exemption is realized by the end of 2006. The new rural policy further jeopardizes the operations of the village committees by cutting off their chief source of finance. The village committees still rely on the tax levied on farmers to build reservoirs and repair the roads. The paralyzed public sector can no longer sustain any initiatives for the common interest as a result of the tax reform. The 2004 statistics suggested that Shanghe and Red Flower had no public revenue whatsoever since neither had any collectively owned enterprises. In general, the weakening of collectivism and the disempowerment of rural administrations foreclose any administrative efforts aimed at improving the local infrastructures and the social welfare.

10.5 The Experiences of Tongnan and their Implications

Despite the short life span of the project, the Internet does serve as an agent of change to induce development to Tongnan County, even though the scale and the depth of the changes are not enough to support the widespread confidence in the new communication technology. The project has brought with it the overall improvement of the telecommunication infrastructure in Tongnan County. More efforts have been made by the county government to build up and renovate the communication and information transmission systems to facilitate the diffusion and application of the Internet and other ICTs employed by the project.

Concomitant with the implementation of the project, cable TV and fixed-line telephone services have been extended to each village in the county. Tongnan gets connected with the national telecommunication network and 5,000 long-distance microwave and optical-fibre links are installed to fasten the transmission of information. The enhanced availability and quality of telecommunication infrastructure helps to foster an environment conducive to the development and use of ICTs. The service providers such as China Mobile and China Unicom have developed 570,000 mobile phone subscribers and Internet cafes have grown from 5 to 153 within the project period.

The information technicians as well as those who have acquired E-literacy through training at the Internet centres help to ease the shortage of technical personnel. With the command of computer and Internet skills, the part time technicians affiliated to the town and county governments take the responsibility for the maintenance and repair of the computer and network facilities of all government institutions. They also provide advice and assistance free of charge to public institutions and private enterprises. These innovators together with the early adopters greatly enhance the technical capacity of the rural county to harness the Internet for development purposes.

The integration of the Internet with the existing communication channels to raise digital awareness and spread advanced farming technologies helps to increase farmers' access to information and knowledge. In addition to conducting information searches for the farmers, the town and village Internet centres also act as information gatekeepers by sifting through the most relevant websites for useful and reliable information. This has to a great extent facilitated farmers' information processing and reduced the level of risks involved in making farming decisions on the basis of online information. The coordination between the town and village Internet centres and the agricultural institutions enable these institutions to be more aware of the development needs of the farmers so that efforts could be made to help farmers come up with optimal solutions.

Despite the fact that only a tiny portion of the rural people at the project sites have experienced tangible benefits from the Internet technology, they may serve as role models and thus bring about a radiating effect among their neighbours and friends who look up to them. This may gradually change rural people's perception about the Internet and its potential role in improving their life.

Last but not the least, the most profound and far-reaching implication that the Internet has on the local development is that on the village school students. Their exposure to the Internet, computers and fax machines may have a positive influence on them. The experiences of E-literacy training and conducting information searches for the adults provide the schoolers with valuable opportunities to get to know the conveniences and efficiency the new ICTs could possibly achieve. If it had not been for the UNDP-MOST project, computers could only exist in the imagination of the youngsters.

11. Pinggu District

11.1 The Profile

Pinggu is a suburban district in the northeast of Beijing Municipality. With an area of 1,075 square kilometres and a population of 400,000, Pinggu administers 16 townships and 275 villages. As one of the major agricultural production bases of Beijing, Pinggu prides itself on the cultivation of peaches. Nearly 30,000 rural households are engaged in peach production in a planting area of 11,200 hectares, with the annual output of fresh peaches reaching 150,000 tons. Pinggu peaches are sold in more than 20 provinces nationwide and exported to a dozen countries such as Russia, Singapore and Mongolia, bringing in an estimated income of U.S.$5 million per year. With rural people making up over 80% of the district population, the annual per capita income of local farmers stood at RMB 6,550 (U.S.$819) in 2004 (PIO, 2005).

Pinggu began to strengthen its ICT infrastructure towards the end of 1998, when the district government set up the Information Office to assist government institutions and local enterprises in using the Internet. In June 2000, the local government initiated the agricultural information network project to connect farmers with the Internet. Taoyuan Village of Zhenluoying Town and Nandingfu Village of Machangying Town were two pilot villages selected to carry out the experiment.

11.2 Taoyuan Village

Taoyuan is a remote mountainous village in the far north of Pinggu. In Taoyuan, 327 residents of 97 rural households live in an area of 3.38 square kilometres. Before 1985, local farmers relied on growing crops to make a living. The poor land produces could hardly help them make ends meet, with a per capita annual income of RMB 200 (U.S.$25).

With the implementation of the household contract responsibility system in 1985, Taoyuan farmers had more choices in terms of planning farm work and distributing farm products. However, as the per capita land allocation was only 500 square metres of hilly land, several farmers started to think of going outside to make money. They first purchased pears from neighbouring villages and stored them over the winter. At the springtime they went to the northeast provinces to sell the pears at a higher price. With the success of the initial trips, a dozen rural households began to emulate the practice. The per capita income of these households jumped to RMB 5,000 (U.S.$625) in 1987.

After ten years of development, Taoyuan has become a village specialized in fruit shipping and marketing. Over 80% of the rural households are now engaged in storing and selling fruits, whose business networks cover Liaoning Province in the north and Guangdong Province in the south. The annual sales of dried and fresh fruits reach 3,200 tons, bringing in an income of RMB 1.8 million (U.S.$225,000). The average earnings of fruit traders have increased to RMB 6,000 (U.S.$750) per year.

Despite the fact that fruit marketing helped Taoyuan farmers get rid of poverty, their modes of selling remained primitive and backward, which were just buying and selling of commodities at different places. Due to lack of market information, fruit traders needed to go out to look for potential buyers. This greatly increased the prices. When failing to find a ready market in time, farmers often ended up bearing the loss caused by rotten fruits. In order to facilitate farmers' access to market information and help them increase the scale of fruit business, the Information Office of Pinggu Government introduced the Internet to Taoyuan in 2001.

With hopes of making Taoyuan the first Internet village in Pinggu, the Information Office employed existing channels of communication to spread the innovative idea of Internet adoption and convince farmers of the benefits of online information to their fruit business. The village loudspeakers broadcast several times a day the successful stories of Internet use by farmers in other regions and door-to-door visits were paid to rural households to answer their queries about the technology. To the promoters' surprise, the Internet diffusion initiative was met with high enthusiasm among local farmers, who were curious about what computers were like and how the Internet might bear relevance to their life.

In view of the scattered rural households in the mountains, the Information Office did not choose to set up an information centre to offer collective Internet access. Instead, it encouraged individual households to buy computers and log onto the Internet at home. Under the financial support of the district government and the village committee, over 30 families bought self-assembled computers at a price of RMB 4,000 (U.S.$500), half the cost of branded products.

Because the local telecommunication operators did not provide the Internet connection service, the Information Office together with one cable TV company made use of the existing cable TV network to set up Internet access for the rural households. A cable modem was installed at the subscriber end to allow a computer to be connected with the Internet. The cable Internet access enabled farmers to enjoy high speed Internet service.

In order to help farmers quickly gain computer and Internet application skills, the Information Office dispatched technical staff to offer them basic E-literacy training. The training sessions were arranged at nights when farmers were less busy. As the technicians were all college graduates, the computer terminologies they used in teaching were far beyond the comprehension of local farmers. A more pragmatic way of teaching was devised to adapt to the learning capability of farmers. Instead of introducing the concepts of "Microsoft Word" and

"Central Processing Unit (CPU)", technicians directly instructed farmers how to access agricultural websites through keying in English letters. This proved effective for the trainees, the majority of whom were in their thirties or forties and had hardly completed junior middle school education. After practising for a month, around 40 farmers learned the basic skills of typing, computer operation and searching information online.

In 2001 the Information Office also helped Taoyuan set up a village website (http://taoyuan.bjpg.gov.cn) to promote their fruit products and facilitate business transactions. The village website covers the history of Taoyuan as well as how the village evolves into one specialized in fruit production over the years. There are also sections promoting local farm produces such as walnuts, pears and peaches. On the website villagers can publicize information about the fruit products for sale as well as contact details. By clicking the online shopping links, potential clients can immediately send their orders over the Internet. The information posting section allows any visitor to post supply and demand information on the website.

The Internet exposed Taoyuan farmers to innovative ways of selling their products. The online communication saved the trouble of travelling great distances for sales and many clients contacted local fruit traders by calling the phone numbers they listed on the website. The farmers also used the Internet to keep track of the fruit prices at the region's wholesale market in order to negotiate for better terms in their transactions.

With an increasing number of sales, Taoyuan farmers set up a Fruit Marketing Cooperative to coordinate their business efforts. In August 2002, a company in Guangzhou placed an online order of 600 tons of peaches and required at least eight tons to be sent to the designated place per day. In order to meet the requirement on time some farmers took charge of collecting peaches at neighbouring villages while others were busy negotiating for freight trains. After two months' hard work, Taoyuan farmers not only successfully fulfilled the order but also established a long-term business relationship with the company.

During the first two years since the Internet was introduced to Taoyuan, the technology was intensively used by local farmers. The village committee covered the Internet subscription costs by paying a monthly flat rate of RMB 50 (U.S.$6.25) for each networked household. Business contacts were established with more than 20 enterprises across the country. Over 1,400 tons of fruits were sold and farmers in the village earned RMB 800,000 (U.S.$100,000) in total (S. Liu, personal communication, August 5, 2005).

The Internet was gradually changing the lives of farmers at the mountainous village. Instead of playing cards or drinking in winter seasons, dozens of farmers preferred to surf the web. They often visited Farmers' Internet School at the website of district government, where they had direct access to agricultural information and farming skills. The children used computers to download educational materials from the websites of elite middle schools and distance learning programmes, which served as a supplement to classroom learning.

The experience of the Internet adoption brought fame to Taoyuan. The district government set Taoyuan as a role model to demonstrate the benefits that the Internet could bring to rural economy. Government officials at various levels came to inspect the results of Internet diffusion and visiting tours were arranged for farmers in neighbouring towns to learn of the advanced practice.

However, farmers' enthusiasm stirred up by the Internet started to die down and the usage of the technology declined in 2004. Among over 30 rural households that adopted the Internet, two thirds sold their computers. During my stay at Taoyuan in August 2005, the Internet access was no longer available because the cable system remained disconnected for months.

One of the reasons behind this decline is that the village stopped paying the monthly subscription fees for the farmers in early 2004. The village head explained that costs for Internet connection became a huge financial burden for the village committee. With no village enterprises, the village did not have the money to pay for the access fees, which totalled RMB18, 000 (U.S.$2,250) per year. Households without computers thought it was unfair to use public funding to benefit only part of the community.

Moreover, the Internet service provided by the cable company was not well maintained. When the cable was down because of bad weather like thunderstorms, it might take months for the cable company to attend to the problem. Most often, the Information Office had to beg the company to repair the network. Due to low Internet subscription rate in the mountainous region, the cable company was less interested in providing Internet services to farmers since it did not generate enough profits.

The village has also been receiving less technical support from the Information Office. One of the farmers, Zhang Shujun, said that previously whenever machines broke down, the Information Office would send technicians to resolve the problem within a day or two. However, with more and more government institutions going online in Pinggu, the Information Office could not offer timely support as before.

Quite a few young adopters at the village complained about the inconveniences caused by the termination of Internet service. As Li Huaiquan put it, he used to monitor the price fluctuations of fruits over agricultural websites and make selling and buying decisions accordingly. During his leisure time he often exchanged emails and played computer games. As a result, the interruptions of Internet access made him quite frustrated. In response to the poor maintenance and lack of technical support, he refused to pay for the monthly subscription fee. Likewise, most of the local adopters distrusted the cable company as a reliable service provider and therefore gave up using the Internet.

Consequently, the intermittent Internet connectivity and lack of timely technical support soon made farmers lose their interest in using computers. When the village committee stopped subsidizing Internet access, the farmers gave up their adoption. They do not think it is cost effective to pay RMB 50 for accessing the Internet because the same amount of money can cover the household monthly expenditure on vegetables. To date, only ten households still keep their computers, which remain idle for most time of the year. Now there is less hype about the Internet village. Taoyuan farmers still depend on the village website for business opportunities and the Information Office updates the web content at farmers' requests from time to time.

11.3 Nandingfu Village

Nandingfu Village is under the direct administration of Machangying Town, situated in the west of Pinggu. The village has 235 rural households and 610 people with about 73 hectares of arable land. Thanks to the robust development of the village economy in recent years, the collective revenue of Nandingfu reached over RMB 4 million (U.S.$500,000) and local farmers' per capita annual income stood at RMB 8,100 (U.S.$1,012) in 2004 (S. Liu, personal communication, August 5, 2005).

Like other villages at Pinggu, Nandingfu practiced the household contract responsibility system in the 1980s and 1990s. With the land redistributed to each household, the family became a basic economic unit instead of the village collective. However, the land system did not work well in Nandingfu. As farmers tended to be only concerned about the productivity of their own fields, disputes arose frequently on the fair use of public resources and daily agricultural activities such as irrigation and harvesting. With the relationship among rural families getting intense, the village committee had to spend most of its time and efforts in settling disputes. Moreover, Nandingfu did not see much improvement in the living standards of local farmers with the adoption of the land system.

In April 2001, the election of the new village committee led to a series of reforms to boost the village economy. The local farmers unanimously agreed that Nandingfu would revert to the collective economy and give up the household responsibility system. The adoption of the collective system allows a score of farmers to cultivate all the fields in the village with the aid of modern farming machineries and at the same time frees up 90% of the labourers (around one third of the village population) from agricultural production. By renting the public land to private enterprises and establishing a clothes factory, the village has managed not only to utilize the surplus manpower but also to greatly increase its collective revenues.

Benefiting from the growing collective economy, Nandingfu becomes way ahead of other rural communities in terms of the development of basic infrastructures and the provision of social welfare. Gardens, clinics, and sports and entertainment centres have been built to improve farmers' living environment. The village has also been subsidizing the household subscription fees for a rural newspaper since 2001. In order to provide farmers with opportunities of using modern communication technologies, Nandingfu decided to promote E-literacy among local farmers at the end of 2002.

The initiative received full support from the Information Office of Pinggu government, which attempted to seek a viable model of diffusing the Internet among rural residents. Given that Nandingfu could afford the potential costs associated with the Internet collectively, the Information Office suggested that an information centre should be set up to offer public Internet access.

With partial financial support from the district government, Nandingfu invested RMB 50,000 (U.S.$6,250) in buying 19 computers and 50 desks and chairs. Of the computers, 15 were put in the information centre and the rest were used by the village committee for accounting and daily management purposes. The information centre is located next to the office of the village committee, which is within walking distance for all the households.

When the information centre was first set up, the village often invited the technicians at the Information Office to deliver lectures. Anyone with an interest was given an opportunity to learn. After a couple of training sessions, around 60 young farmers acquired skills to use the Internet. A female farmer named Bao Suohong became a part time manager of the information centre.

Unlike other E-centres, the information centre at Nandingfu does not provide information and communication services to the villagers. Instead, it is used as an E-literacy training centre by the town government. Because it is the only one of its kind in town, the centre has been extensively used to promote E-literacy among farmers at surrounding villages under the direct administration of Machangying Town.

The chief organizers of training courses are often various branches of the town government, such as agricultural or judicial sectors, women's association, and youth organization, etc. Farmers enrol in the courses on a voluntary basis without paying tuition fees. A typical training course usually lasts three months including both formal teaching and hands-on drills.

Teachers from agricultural colleges in Beijing are often invited to deliver lectures. At the beginning of the training courses, teaching materials as well as pamphlets listing the frequently used agricultural and government websites are provided to facilitate the learning process. There are elementary and advanced courses to cater to different levels of knowledge. The most often taught content for beginners includes basic computer operations, information searching, saving files, and sending or receiving emails, whereas advanced courses often deal with webpage design and the use of Microsoft Office software. Farmers who pass the test at the end of the courses are conferred with a certificate.

The manager of the information centre, Bao, said that most of the learners were 25 to 40 years old and about 80% were females because they tended to be less busy than men and many wanted to learn so that they could educate their children. With E-literacy skills learned at the training courses, some farmers have set up their own businesses by offering photocopying and typing services while a few have bought computers and started to surf the Internet at home.

According to Bao, though computers and the Internet are unlikely to play a major role in rural people's life at the present stage, the E-literacy training helps to create awareness among farmers about the potentials of the technology and provides them with the opportunity to learn the basic skills of using the Internet. Otherwise farmers would be easily bypassed by the technology.

Since the establishment of the information centre, all the costs of hardware maintenance and Internet access have been covered by the village committee. The centre had been connected to the Internet via a phone-dialling mode until the district government invested RMB10, 000 (U.S.$1,250) to extend broadband service to the village in 2004. The village head, Li Guang'en, said that the Internet subscription fees were around RMB 4,000 (U.S.$500) per year. Even though the village currently has no difficulty in shouldering the costs, Li still hopes that he can negotiate rates on a collective basis with the local ISP so that the village may avoid paying access fees according to the number of computers. In order to widely diffuse computer and Internet application skills among rural people, Pinggu government has given Nandingfu 31 new computers and encouraged it to set up another information centre in late 2005.

11.4 The Experiences of Pinggu and their Implications

Following the experimental project carried out at two pilot villages, Pinggu has set up 12 information centres targeting at farmers at half of its towns within four years.

The information centres have trained 5,600 information technicians at the district, township and village levels and 38,000 farmers. By the end of 2005, 5,773 fibre kilometres of optic cables had been installed to link up every village in the district (CRTDC, 2005).

The diffusion of the Internet in Pinggu is characterized by a strong government involvement. The Information Office of Pinggu government serves as a change agency to promote the Internet diffusion and adoption among rural people. Farmers at Pinggu would be like farmers in other parts of the country lagging behind the urbanites in terms of communication and information exchange, if it were not for the efforts made by the local government to build telecommunication infrastructure and provide collective access for the rural residents.

The two pilot villages, Taoyuan and Nandingfu, have shown contrasting experiences even though the Internet diffusion at these two places was initiated by the same change agency. Whereas Taoyuan farmers adopt the Internet mainly on the individual basis, Nandingfu farmers share public computer and network facilities at the information centre. Taoyuan farmers have more opportunities of accessing the Internet and thereby experience more tangible benefits of the technology. However, to most young farmers at Nandingfu, their adoption only stays at the stage of awareness and skill acquisition. The Internet is still unlikely to impact their lives much because of their limited use of the technology.

Despite the different ways in which the Internet has been embraced by the two rural communities, the utilization of the technology is greatly influenced by its planner. As farmers depend on the financial and technical assistance provided by the government to be introduced to the Internet, their use is guided and purposive from the very beginning. In the case of Taoyuan, farmers took the Internet as an efficient tool of searching price information and establishing business contacts. E-literacy is perceived as an essential skill for a modern citizen by farmers undergoing computer training at the information centre of Nandingfu.

Pinggu's experiences also suggest that ready access to the Internet does not necessarily guarantee sustained usage of the technology by farmers. The adoption and continuous use of the Internet are determined not only by individual farmers' economic status and educational level, but also by the strength of the village economy and the commitment of local ISPs. Due to the limited Internet penetration in rural areas, farmers are unlikely to attract commercial support from both the service and content providers. At Taoyuan, the poor Internet access provided by the local cable company soon dampened farmers' enthusiasm for the Internet. Because there are no existing state policies to subsidize the rural telecommunication operators, the high

Internet access fees become another important factor constraining the expansion and sustainability of the ongoing Internet initiatives.

With regard to the social and economic impact of the Internet, the case of Taoyuan has clearly demonstrated the value of the Internet in increasing farmers' access to market information and broadening their business perspectives. At Nandingfu, the town government mainly relies on the existing computer and network facilities to provide E-literacy training and therefore downplays the role of the Internet in facilitating the flow of information and knowledge. Though little information is gained about farmers' direct use of the Internet, the large scale E-literacy training carried out in Pinggu can raise farmers' awareness of new ICTs and equip them with the basic computer operation skills, which may greatly facilitate their future adoption and use of the Internet.

12. Shouguang City and Huanglou Town

12.1 The Profiles

Shouguang City and Huanglou Town are both under the administration of Weifang City, Shandong Province. Situated in the monsoon area of the North Temperate Zone, Weifang is characterised by striking seasonal changes with an annual average temperature of 12.3 °C, a frost-free period of about 200 days, and an average precipitation of over 700 mm. All these provide exceptionally favourable conditions for the local agriculture.

Shouguang City is situated on the southern coast of Laizhou Bay in Shandong Peninsula. It was upgraded from a county to a county-level city in June 1993. The city covers a total area of 2,180 square kilometres with a population of 1.08 million, 82% of whom are rural residents. The city administrates 21 townships and 976 villages ("Shouguang Gaimao", 2005).

In 2003, Shouguang ranked 53rd among China's 100 top counties in terms of overall rural economic strength. The 2004 per capita income for the rural people stood at RMB 5,016 (U.S.$627), 60% of which came from growing green-house vegetables ("Shouguangshi Gongzuo", 2005). A farmer named Wang Leyi pioneered the "Warm Vegetable House for Winter" (the local version of an inexpensive green-house) by using tunnels constructed of bamboos, plastic sheets and covering matting in 1989. The technology rapidly spread and is now extensively used by farmers in Shouguang.

After years of promoting standardized production and popularizing new agricultural technologies, Shouguang has developed into China's largest vegetable production base, covering 53,000 hectares (almost 60% of its arable land). The city has successfully held its International Vegetable Science and Technology Fair for six consecutive years. As the national vegetable trading centre, Shouguang vegetable market boasts of an annual trade volume of 350 million US dollars. The vegetables are sold to more than 200 cities in China and exported to Russia, Mongolia, Japan and many other countries and regions ("Shouguangshi Gongzuo", 2005).

To the southwest of Shouguang, about 30 kilometres away is Qingzhou City, where Huanglou Town is located. While Shouguang is renowned for its greenhouse vegetables, Huanglou gains fame through its flower production. In Huanglou, about 39,000 people live in 50 villages within an area of 52 square kilometres ("Jiceng Nianbao", 2005). The agricultural population makes up 84% of the town population and their per capita income reached RMB 4,260 (about U.S.$532.5) in 2004.

As the biggest flower production and distribution centre to the north of the Yangze River, Huanglou grows flowers on 1,000 hectares of land, almost one third of its arable land. Over 35% of the town's yearly revenue comes from flower sales, which amounted to RMB 2.1 million (U.S.$262,500) in 2004. The prosperity of the flower industry gives rise to nearly a hundred rural enterprises and more than 3,800 households (40% of the rural households) engage in flower production and marketing. Huanglou has been hosting the annual Shandong Flower Exposition since 2001, which has made transactions worth over RMB 1 billion (U.S.$125 million) accumulatively and attracted flower gardeners and traders from over twenty provinces as well as a dozen countries and regions in the world ("Huanglouzhen Qingkuang", 2005).

12.2 The Diffusion of the Internet

Thanks to the overall social and economic development, the telecommunication infrastructures of Shouguang and Huanglou have undergone rapid improvement in recent years. In Shouguang, TV and cable TV penetration rates reached 95% and 70% in 2004. Fixed-line telephones are extended to every village and half of the villages are connected to the Internet (MOST, 2005b). In the case of Huanglou, all rural households have access to cable TV, and fixed-line and mobile phone subscription rates stood at 66% and 32% of the households in 2004 ("Jiceng Nianbao", 2005).

Though Shouguang and Huanglou are endowed with relatively developed rural infrastructures to sustain the diffusion and adoption of the Internet, the technology has not been widely used among local farmers. Even though a few rural families do have computers, they are mainly for the children's education. It is extremely rare for local farmers to subscribe to Internet services for other purposes. The findings at these two places indicate that several factors have contributed to farmers' low Internet adoption.

First of all, the cost of computer hardware and network facilities is the equivalent of a whole year's income for an average rural household. Farmers do not think Internet adoption can be cost-effective, as the investment does not necessarily promise financial returns. This explains why it is common for many rural households of Shouguang to own trucks instead of computers.

Second, farmers' education level does not necessarily grow with the increase of income. Most farmers have not finished their junior middle school education, which makes it hard for them to learn computer skills. In particular, without knowing the Hanyu Pinyin Phonetic System, farmers do not even know how to key in Chinese characters. When asked about how he perceived the Internet, a farmer from Huanglou offered a vivid analogy, "To us farmers, a computer is no different from an aircraft carrier because neither has a bearing on our life".

Third, the Internet does not seem to bring direct benefit to daily agricultural productions and therefore farmers have fewer needs for the technology. Both Shouguang and Huanglou have already become famous markets with thousands of traders coming over every year. As small-scaled household productions dominate the local agriculture, farmers do not have difficulty in selling their products, which are often quite limited in quantity. They may just wait for buyers to knock at the doors or sell at the local markets.

Moreover, due to a lack of relevant content, the Internet cannot compete with the existing communication channels in the rural communities. With wide adoption of new technologies such as soilless and greenhouse cultivation and the transition from traditional to precision farming, farmers need to grasp enough knowledge to adjust to the changes and tackle production problems. Due to a lack of research skills and expertise, the available agricultural websites can hardly fill the information needs. Instead, the local farmers have to turn to alternative information channels.

Take Huanglou as an example, the annual flower expositions provide farmers with wider opportunities of getting exposed to the latest development in flower cultivation. During the expositions many foreign seedling companies offer lectures to facilitate farmers' adoption of their products. Through the lectures and distributed materials, farmers are able to learn of advanced farming technologies. In addition, most farmers have a habit of subscribing to *China Flower* (a newspaper) and *China Gardening* (a magazine) to obtain updated information on farming skills. As the two publications often carry articles written by successful flower growers based on their experiences, they enjoy a high readership among the local farmers.

Though the Internet is not well received by the majority of the local farmers, several rural pioneers have started to adopt the Internet since the technology was newly available in rural areas. These innovators use the Internet mainly as a tool to promote their enterprises and facilitate business transactions. The next section documents the Internet adoption and usage of three rural pioneers, who enjoy great prestige due to their leading positions in the local development.

12.3 The Stories of Three Rural Entrepreneurs

12.3.1 The "Vegetable Doctor" Zhang Yanxiang

Zhang Yanxiang, 34 years old, began to use the greenhouse technology to grow vegetables after graduation from a junior middle school in 1993, just as thousands of other Shouguang farmers did. He had expected a bumper harvest since the tomatoes grew extremely well under his careful management. However, a plant disease started to spread rapidly in his greenhouses. Though Zhang searched desperately for proper treatment, his efforts turned out to be fruitless as few people around knew about plant disease control.

Ever since the frustrating experience, Zhang has engaged in studying plant diseases and pesticide formulations. Despite limited formal schooling, Zhang studied by himself the textbooks on plant biology and pathology used in agricultural universities. After five years of study and experiment, Zhang invented two kinds of pesticides to prevent and control vegetable bacteria and tomato late plight, both of which have been filed for patent. In order to let his inventions benefit the largest number of farmers possible, Zhang transferred the technologies to a local pesticide factory for mass production.

In 2002 Zhang set up the Centre for Prevention and Control of Vegetable Diseases and Pests in Luocheng District of Shouguang City to propagate knowledge and technologies among farmers and increase their abilities of coping with vegetable diseases. With his expertise as well as the aid of high-tech equipment, Zhang has become a popular "vegetable doctor". During high farming seasons the centre is often crowded with hundreds of anxious farmers, many of whom travel over long distances to seek advice on their diseased vegetables. After diagnosing the problems, Zhang also "prescribes" pesticides and informs farmers of the right ways of application. Most of the pesticides can be bought from the centre at reasonable prices. Zhang's persistent efforts over the years have helped many local farmers avoid severe damages caused by diseases and pests.

With the rapid extension of greenhouse farming and increasing vegetable planting areas in Shouguang, the local production is plagued with higher incidences of diseases and pests than before. The plant protection institutes can not offer much help as they are not experienced in managing diseases and insects attacking greenhouse vegetables. Zhang hoped that he could search relevant information and solicit advice from agricultural experts in other parts of China via the Internet.

Zhang bought his first computer for RMB 9,000 in 1998. At that time dial-up was the only Internet access method and the service fee was very expensive. It often cost Zhang over 300 Yuan to cover the monthly phone bills. Zhang said he did not have many difficulties in acquiring Internet application skills except that he spent quite some time learning the stroke approach to enter Chinese characters. Zhang often went to the websites of agricultural universities and plant protection institutes to search for the latest findings on disease control.

However, Zhang did not find much useful information over these websites. He soon realized that because greenhouse farming was not adopted in other rural areas on such a large scale as in Shouguang, the occurrences of vegetable diseases were less frequent elsewhere. In view of this, Zhang decided to set up a website where he could publicize the discovered cases of diseases over the Internet and share information with other people. Shouguang Yanxiang Vegetable Disease and Pest Control website (http://www.bchfz.com) was established in 2001. A webpage design company in Weifang City was hired to create and maintain the website, while a university graduate specialized in plant pathology is responsible for updating the contents.

One of the main functions of the website is to post online the prevalent vegetable diseases so far discovered in Shouguang. Among 214 kinds of vegetable diseases, half of them belong to complicated cases. The diseases are classified into ten groups according to the vegetables they often attack, such as cucumbers, eggplants, tomatoes, peppers, and cabbages, etc. By clicking the name of each disease, users may find a detailed description of symptoms and recommended pesticide formulas. The pictures of the diseased vegetables are also placed right beside the texts to facilitate users' self diagnosis. The webpage seems less frequently visited, with the total number of clicks for each disease category ranging from several dozen to over two hundred.

Besides this, the website has a plant protection forum where users can share their experiences in controlling vegetable diseases. In the forum a score of visitors posted the diseases online and each of such inquiries was followed by four to five postings either from Zhang and his assistants or other visitors to offer various methods of treatment. Zhang said that the forum was meant to facilitate knowledge sharing between farmers and agricultural experts. However, many visitors prefer to call the centre to seek advice rather than to discuss their problems online.

Though Zhang expects to use the website to spread disease control knowledge to the widest range of farmers possible, the frequent visitors are from plant research institutes and agricultural universities because few farmers have regular access to the Internet. Many agricultural experts outside Shouguang know of Zhang through his website and they often exchange e-mails to discuss complicated cases. Most often, feedbacks come through phone calls. From time to time, Zhang receives calls from certain plant protection institutes informing him of alternative ways of disease treatment other than those offered at his website, which have demonstrated satisfying results so far in the local application.

Other than the establishment of the website, Zhang used the Internet to develop a telediagnosis system. When farmers call over phones for help, they often fail to give a complete and exact description of the symptoms of the diseased vegetables, which makes it difficult for Zhang to make an accurate diagnosis. Depending on a system similar to cyber conferences except with images of high resolution, Zhang is able to observe the diseased vegetables under the digital camera and make a quick diagnosis. Though still at a primitive stage, Zhang hopes that the telediagnosis system can offer timely help to those farmers living in provinces far away from Shandong.

Though there is a great need for knowledge on disease control among farmers, the setting up of the website does not help much in this respect. According to Zhang, the local farmers are unlikely to buy a computer and log onto the Internet to search for cures to their diseased vegetables because this is not a cost-effective way. Instead, they would rather take the vegetable samples to his centre or plant research institutes for advice. Therefore, Zhang feels the need to seek more practical ways of disseminating agricultural technologies among Shouguang farmers. His experiences suggest that providing agricultural training and establishing farmer associations are among the effective ways of meeting the purpose.

Every year for one month in busy farming seasons, Zhang travels to various villages in Shouguang to deliver lectures on greenhouse management. Many of such visits are arranged at the request of farmers who have been to Zhang's disease control centre. Zhang can easily get his ideas across by using farmers' language to impart new concepts:

> I can feel the uttermost sense of achievement when being surrounded by listeners. They are in desperate need of knowledge and information. I was once a farmer myself and I can fully understand the frustrations and setbacks that local farmers experience when their vegetables are attacked by diseases or pests. Most often, they get ten answers from ten different people.

This may explain why Zhang's training sessions enjoy such popularity among his rural audiences.

Another effective way of extending the knowledge of greenhouse management is through the formation of farmer societies. Encouraged and subsidized by Shouguang government, Zhang organized a farmer society on vegetable disease prevention and control in 2004. As a non-profit organization, the society first enlists skilled farmers with an open mind to science and technology and provides them free training on greenhouse management. Then these members are responsible for disseminating the knowledge among other farmers and attracting more people to join the society. The farmer society is also devoted to promoting standardized production by coordinating the selection of seeds, pesticides and fertilizers among its members so that they can collectively stand up to the challenges posed by the market economy.

12.3.2 The Grape Master Huang Rongming

At Huangjiazhuang Village of Hou Town in Shouguang City lives a 72 years old man named Huang Rongming, who is an expert in growing grapes. In May 1991 Huang succeeded in cultivating off-season grapes with his greenhouse of 315 square metres producing 867 kg of grapes. By advancing the grape harvest time, Huang made a profit of RMB 26,800 (U.S.$3,350).

Motivated by the initial success, Huang engaged in breeding new grape species that are suitable for growing inside a greenhouse. After several years of experiment, Huang discovered three grape cultivation technologies that are patented for distribution around the country. Thanks to Huang's efforts, the greenhouse grapes ripe 120 days earlier than those produced outdoors and the output may reach 2,000 to 3,000 kg per mu (about 666 square meters).

Now Huang's grape yard covers 26 hectares where 14 greenhouses have been built for production, research as well as seedling nursery. Huang has registered a trademark for his grapes with Rongming as the brand name and his grape yard has been certified as one of the pollution-free agricultural production bases of Shandong Province by China's Ministry of Agriculture. As a rural entrepreneur with an annual income of over RMB 0.5 million (U.S.$62,500), Huang's success is attributable to his never-ending pursuit of knowledge. Despite his age, Huang is among the first in Shouguang to sell his grapes over the Internet.

At a retiring age, Huang has an inquisitive mind of the young. His small office of less than 12 square meters is crammed with all kinds of ICT appliances. Initially Huang took pictures of grapes at various stages of growth. Later on he improved the demonstration by employing video cameras, colour TVs and video cassette recorders (VCRs).

With the advent of the digital age, Huang uses video compact discs (VCDs) to better promote his technologies among farmers.

Huang reads 27 newspapers and over 30 journals, most of which are concerned with agricultural technologies, especially those for fruit cultivation. The annual subscription fees cost more than RMB 1,000 (U.S.$125). It is from these journals that Huang learned about the Internet. Upon reaching the age of 66, Huang bought a computer for RMB 12,000 (U.S.$1,500) and learned the basic skills from several college graduates working at his greenhouses. Since then Huang has been accessing the Internet through a phone-dialling mode because the broadband service is yet to reach the village.

After setting up a personal webpage in 1999, Huang invested RMB 50,000 (U.S.$6,250) to create China Rongming Grape Website (http://www.chinarongming.com) four years later. Technicians of a web company based in Weifang City visit Huang's office every week to update the website and maintain the hardware. At the website visitors may find detailed information about the top grape species as well as information on disease prevention and control. The online shopping function allows grape traders to place orders or make payments over the Internet. The email accounts and phone numbers of Huang's company are also listed on the website to facilitate communication with potential business partners.

According to Huang, the Internet has proved to be an efficient tool in increasing sales. Now over 70% of the transactions are achieved through business contacts established via the Internet. The website makes Rongming grapes widely known both at home and abroad. Two professors of Florida Grape Grower's Association read about Huang over the web. Through exchanging emails they quickly became good friends. In February 2002 the American professors travelled all the way to Shouguang to visit the Chinese grape master. They brought the old man three grape species developed in the U.S. and in return Huang gave them Rongming Grapes No.2, No.3 and No.4 as gifts.

In addition to introducing Rongming grapes online, Huang and his assistants also go to other main agricultural websites to publicize product information. Posting information online helps them save the cost of advertising since most often this can be done at no charge. Everyday inquiries and orders come through phone calls or emails. Over 90% of grape traders and growers from elsewhere know of Rongming grapes through the Internet. After 14 years of persistent efforts Huang's technologies have spread all over China with the grape planting areas covering more than 2,600 hectares. Rongming grapes are sold at supermarkets in a score of cities and also exported to other countries.

Despite his personal success, Huang spares no efforts in helping the local farmers get rich. In order to quickly spread grape cultivation technologies, Huang gives the villagers grape seedlings and teaches them the necessary skills free of charge. Under Huang's guidance, all the 109 households in the village are engaged in grape growing. Now the village produces over 370,000 kg of grapes and two million quality seedlings every year, which brings an extra income of RMB 8,000 (U.S.$1,000) per person. The development of Huangjiazhuang Village has had a ripple effect among its neighbouring areas. With more and more farmers cultivating greenhouse grapes, Hou Town has become one of the largest grape production bases in the country, which devotes 133 hectares of land to grape cultivation. Huang also uses the Internet to connect local farmers with the markets outside. He often helps farmers post product information online and establish business contacts with potential buyers. With the help of Huang and his website, the greenhouse grapes are often booked up even before the harvest time.

12.3.3 The Flower Grower Zhang Jizhong

Zhang Jizhong, 39 years old, is one of the most successful flower growers in Huanglou Town. After graduating from university, Zhang became a middle school teacher. Four years later, he quitted his job and started to try other jobs in the hope of making big money. He had been engaged in building roads and running a restaurant before setting up his own enterprise of flower cultivation. There is some ambiguity in calling Zhang a "farmer". If judged by the household registration system, he is an urban resident. However, Zhang considers himself a farmer because he makes a living by growing flowers.

Nevertheless, Zhang is by no means an ordinary farmer. In 2003 he invested RMB1.5 million (U.S.$187,500) in building greenhouses with a total space of more than 2,000 square meters, which utilized such advanced technologies as drip irrigation, weatherboarding, electric curtain rolling and computerized temperature control. Zhang's flower company annually produces nearly one million flower seedlings and 300,000 pieces of red azalea, cyclamen, and cymbidium, etc.

In 1999 Zhang became aware of the Internet through his friends working in the IT sector in Beijing. At the flower market where Zhang worked, he happened to know a trader who sold flowers over the Internet. Zhang went to his store every day to see how the transactions were made. From the trader he knew that with an increasing demand for fresh flowers, many government institutions preferred to place orders online. They often approached the trader to make purchases because he posted his products over the Internet.

Zhang bought a computer and attended a seven-day workshop for E-literacy training. Then he set up a website with the help of a webpage design company at a cost of RMB 7,000 (U.S.$875). The technicians of the company come to Zhang's office twice a month to update the content of his web pages. Now Zhang have two computers, with one in his company and the other at home.

Logging onto the web has become a daily ritual for Zhang. Every day Zhang goes to the company website to read and reply to messages left by visitors. To Zhang and other big flower growers at Huanglou, selling flowers over the Internet has become an alternative way for business transactions. According to the statistics provided by the Flower Office of the township government, the annual volume of online transactions may reach over RMB1 million (U.S.$125,000).

However, the Internet is only one of the links in the selling process. After seeing the pictures of flower samples on the company website, potential clients often make inquiries through phone calls. If interested, they would visit Zhang's greenhouses to select and check the quality of the flowers before placing orders. Therefore, the Internet acts as an additional information and communication channel for flower growers to establish business contacts.

As compared to business deals achieved via traditional means of communication such as annual flower fairs or sales representatives based in flower markets outside Huanglou, the Internet transactions account for one-tenth of Zhang's total volume of transactions, which is approximately RMB150,000 (U.S.$18,750) per year. Zhang believes online businesses have just started to evolve and it still takes time to see how much the Internet can do because there are currently few users in the agricultural sector.

Zhang often visits the websites of other flower suppliers to monitor their new products and identify market trends. The Internet also serves as a virtual meeting place, where Zhang exchanges information with sales representatives and old clients outside Huanglou. He always uses QQ messenger system to catch business partners online. They either have a text chat or talk to each other over the web.

Zhang has acquired much practical knowledge during the process of using the Internet. He compares establishing a website to opening a store at the market. Zhang said, "The store must be placed in a noticeable position so that customers can easily catch it when shopping around." Initially Zhang used "Seedling Company" as the domain name of his website. Later he found when clients searched the Internet, they preferred to key in names of the flowers rather than the word "seedling". Therefore, he had to change the domain name to ensure that his company could be easily found online.

Though Zhang has been using computers for almost five years, he still meets difficulties from time to time. Due to lack of formal training, he often gets stuck during operation. Zhang thus described his experience, "Even the simplest function can make me frustrated. Sometimes the machine almost drives me crazy when I cannot figure out what is wrong. Though I tried all keys on the keyboard and it still did not work! But this seems to be a piece of cake to the youngsters working at my company. They often solve my problems with just a few strokes."

Zhang thinks that the benefits of the Internet are very limited even though it proves useful in attracting business partners. His flowers can be sold anyway, either via the Internet or the traditional means of transactions. The Internet does not add extra value to the products. Moreover, when online clients come to see his flowers, they may visit other flower growers at Huanglou as well. Sometimes the clients ended up striking bargains with his neighbours because lower prices were offered. Though the advantages brought by the Internet are not obvious at present stage, Zhang is confident early adoption will help him take the lead in local flower businesses. He anticipates that he will be among the first to glean benefits when the network grows bigger and more mature.

12.4 The Rise of Rural Entrepreneurs and its Implications

The increase of wealth has not propelled a large number of farmers at Shouguang and Huanglou to adopt the Internet even though they can afford computers and network facilities. Instead, the Internet adoption and usage are closely connected with rural entrepreneurs and their private enterprises. The experiences of the "vegetable doctor", the grape master and the flower grower are just the epitome of the rise of thousands of private enterprises in rural China.

Statistics released by the Research Association on the Private Economy show that by June, 2004 the number of private enterprises rose to 3.44 million with 47.14 million employees. As a new booster of economic growth, the production value of the private sector reached RMB 2 trillion (U.S.$241 billion) in 2003, accounting for one third of China's economy. This indicates that the development tendency of private enterprises has great implications for the country's overall sustainable development (EPRCI, 2005).

The same trend is sweeping rural China, especially coastal areas rich in natural resources and with a highly developed agriculture. Take Shouguang for an example, much of its social and economic progress in recent years owes to the rapid extension of greenhouse cultivation pioneered by a local farmer. With a great reliance on science and technology and a strong market orientation, the new rural elites have not only made personal achievement but also contributed to the development of the local economy.

The three rural entrepreneurs included in the study have started to experiment with the Internet long before their peers, especially at a time when the advantages of the technology had not fully demonstrated themselves. By setting up a website on vegetable disease and control, Zhang is able to timely publicize research findings and facilitate communication between his centre and agricultural institutions. Huang uses the Internet as an efficient tool to sell and promote his grapes. Likewise, the flower grower Zhang is also testing the innovative ideas of selling flowers online.

On account of their favourable attitudes towards technologies and relatively strong command of financial resources, the venturesome rural entrepreneurs are more likely to take the risks associated with Internet adoption. They act as gatekeepers when imparting innovative information carried over the Internet to rural people nearby. Their adoption behaviours may influence how people perceive the Internet and let the benefits derived from the Internet trickle down to those non-adopters. For instance, Huang's grape website is not serving his company alone but also helps to negotiate business deals for neighbouring grape farmers.

Even though the experiences of the three entrepreneurs cannot be generalized to other innovators in Shouguang and Huanglou, they are nevertheless typical in terms of their taking the lead in experimenting and adopting new technologies. What the three innovators share in common are their persistent quests for knowledge, pro-technology attitudes, and pioneering spirit. The flower grower Zhang explained that it was most likely that he would gain an edge over other local flower growers when more and more transactions were conducted online. The vegetable doctor set up the website to publicize knowledge on disease control and engage in developing a telediganosis system since the existing websites failed to provide useful information in this respect.

As far as Shouguang and Huanglou are concerned, the potentials that the Internet has in bettering the lives of farmers in general are not so evident. Affordability does not seem to be the overriding factor in determining Internet adoption among local farmers. Instead, their adoption decisions are much more need-driven, as a result of a careful deliberation about cost-effectiveness involved with the investment.

On one hand, the Internet fails to prove its efficacy in bringing financial returns. Online communication did not help to increase the value of flowers. Because flowers can be easily sold through the existing business channels such as annual fairs, farmers don't have a need for the Internet. On the other hand, the Internet has given way to other popular information channels due to a lack of content with direct relevance to local farming. The widespread practice of specialized operations like greenhouse farming renders the agriculture at Shouguang and Huanglou more developed than that of other rural places. The adoption of new farming practices has generated stronger information needs for advanced farming technologies, which can be hardly met by existing agricultural websites full of common sense knowledge. The increasing information needs contribute to the popularity of Zhang's lectures among vegetable farmers and flower growers' heavy reliance on specialized publications for updated information and knowledge on gardening. Moreover, farmers' low education level has further distanced them from accessing the technology.

The limited diffusion of the Internet among rural people keeps the technology from bringing tangible benefits to its adopters. According to the "Metcalfe Effect", the value of the network to other Internet users tends to grow with each additional user (Best & Maclay, 2002). The fewer parties go online to carry out transactions, the fewer benefits rural adopters can gain from the Internet. This tells us why the flower grower Zhang did not gain an edge over non-adopters in terms of using the Internet to improve flower sales. In the long run, the impact of the Internet will not get full-blown until a certain number of people connected with rural productions adopt the technology.

13. Discussion and Conclusions

13.1 The Process of Internet Diffusion and Rural Development

According to the three major E-readiness indexes TAI, NRI and DAI, China is a country with middle level development in terms of its readiness and capacity to harness the opportunities offered by ICTs and benefit from networked technologies (ITU, 2003; UNDP, 2001a; WEF, 2003). In a developing country like China, the Internet, whose diffusion requires large investment and advanced knowledge, is yet to be widely diffused in the cities, let alone the rural areas where the deployment of the Internet is handicapped not only by the underdeveloped economy but also by the lack of telecommunication infrastructure and computer skills. In recent years, despite the huge investment made by the central government in telecommunication infrastructures to support the deployment of the Internet, the rapid diffusion of the technology has only taken place in urban centres, largely bypassing the vast rural areas where the majority of the country's population reside.

Given the huge socioeconomic divide between the urban and rural areas, it is likely that the diffusion of the Internet would follow the extension pattern of old ICTs like radio and television, except that the Internet is mainly driven by commercial forces whereas the electronic broadcast media are strictly controlled and managed by the government. On the average, the urbanites can always access communication technologies decades earlier than the rural dwellers.

With a massive population and vast territory, China has spent almost 50 years building a national broadcasting network covering over 90% of the population. Rural people are far lagging behind their urban counterparts in terms of technical access to broadcast services as well as household ownership of radio and television sets. Therefore, China is still at the stage of spreading the benefits of old innovations to the entire population and realizing the last mile connectivity of electronic broadcasting. At present, government efforts are increasingly devoted to extending radio and television coverage to the last 10% of rural people at remote and mountainous regions and improving the broadcast service quality for areas already reached by broadcast signals.

Considering the overall development level of the country as well as the underdeveloped rural economy and inadequate physical infrastructure in particular, the majority of 757 million rural people will continue to be marginalized at the periphery of digital communication networks and there are years to come before the benefits of the Internet technology can trickle down to the rural areas in China. The pioneering Internet projects offer enlightening glimpses of the experiences among Chinese farmers to use the Internet for improving their living standards. While the Internet diffusion initiatives at the five rural places may not be universally applicable to other parts of rural China, the experiences of making use of the Internet for farming and marketing may shed light on what could be anticipated should the Internet reach the rural areas on a wide scale.

Chapter 8 to Chapter 12 provide a detailed description of various ways the Internet technology has been embraced by the rural people and the social changes resulting from their Internet adoption and usage. Of the five rural places, Internet diffusion at Yellow Sheep River, Jinta, Tongnan and Pinggu is as a result of deliberate and purposive organizational intervention, whereas rural entrepreneurs at Shouguang and Huanglou have adopted and used the Internet mainly through individual initiatives. Hence, the five rural settings may be categorized into two groups (Internet diffusion with intervention and without) and each is separately addressed to facilitate the discussion hereinafter.

In contrast with Internet access in urban settings where the adoption of the Internet merely involves the initial purchase of hardware and the acquisition of certain level of technical competence (the infrastructure is already in place), the Internet or online information needs to take multiple steps to reach its rural users. On the basis of the commonalities shown by the experiences of the five rural places, the Internet diffusion process is segmented into Internet access and extension efforts, Internet adoption, Internet usage, the implications of the Internet on rural development, and the issue of sustainability to discuss the dominant features as well as the key social factors that shape and condition the process.

13.1.1 Internet Access and Extension Efforts

The experiences of the five rural places with varying geographical conditions and socioeconomic development levels suggest *two* models of Internet diffusion – Internet diffusion under *organizational intervention* vs. Internet diffusion under *individual initiatives*. Generally speaking, the level of telecommunication infrastructure as well as the strength of the rural economy greatly conditions the diffusion patterns and extension efforts made to spread the technology.

The experiences of Yellow Sheep River, Jinta, Tongnan and Pinggu show that the existing economy and infrastructure of the rural areas can hardly sustain the adoption of an advanced technology as the Internet, whose diffusion at these rural places is by no means a natural development. While farmers may have individual choices in selecting the traditional means of communication for information about the market, scientific farming, health and other issues, the Internet as an advanced technology could only be introduced to them through either a government initiative as in the cases of Jinta, Tongnan, and Pinggu or an external investment as in the case of Yellow Sheep River.

Situated in the underdeveloped western regions, Yellow Sheep River and Tongnan have been much constrained by the unfavourable natural conditions, inconvenient transportation, and poor infrastructures. Although the majority of the rural people at these two places have surpassed the poverty threshold, they remain relatively impoverished. Most agricultural productions are primarily devoted to meeting the basic living needs. The low level of teledensity and poor broadcast services at the project sites also suggest that the rural communities are far from being ready for the natural diffusion of the digital technologies. Even for better off rural areas endowed with rich natural resources like Jinta and Pinggu, the inadequate rural telecommunication infrastructure and the widespread computer illiteracy have also hindered local people from harnessing the Internet for development purposes.

In respect of Internet diffusion at the four rural places that is characterized by organizational intervention, the *change agencies* play an essential role in extending the technology to the rural people. The *rationales* behind the introduction of the Internet greatly determine the *ways of Internet access*, the *extension efforts* taken to encourage farmers to adopt and use the technology, and the *direction of social changes*.

Whereas the Internet was introduced to Yellow Sheep River under the initiative of a private company, the diffusion of the Internet in Jinta and Pinggu was funded and supported by the local governments. Tongnan stood out from the group as its Internet project was jointly funded by UNDP, the MOST and the county government. Despite the varying approaches taken to promote the diffusion of the Internet, the change agencies, be they international donors, private companies or local governments, hoped to harness the potential of the Internet and help the rural places experience rapid economic growth and rising living standards. In Yellow Sheep River and Tongnan, the Internet was perceived as a potential tool to alleviate poverty while the local governments of Jinta and Pinggu wished to assist farmers in making informed decisions by increasing their access to online information.

At the four places, great efforts had been made to overcome the *cost* and *knowledge* barriers. The change agencies in Yellow Sheep River, Tongnan and Pinggu (Nandingfu Village) selected the collective access model to spread E-literacy and cater to the information requests of the farmers. The information centres allowed the rural people to use the Internet on a shared basis, which greatly reduced the per capita cost of Internet access. Moreover, information technicians carried out trainings on computer and Internet application skills, and helped farmers search agricultural and market information, and advertise their farm products online.

Unlike the other three places, Jinta did not provide the local people with direct Internet access. Instead, the *multiple-step flow model of communication* was utilized by the county government to disseminate online information to the majority of the rural households. The online paper was printed to cross the barrier of technical incompetence. Thereby Jinta farmers could access information from the Internet without actually surfing the net themselves. The distribution of the newspaper also helped the local government circumvent the high costs associated with Internet access so that updated online information could reach the widest range of rural population in a most cost-effective way.

Besides providing rural people with access to the Internet or online information, the change agencies developed various extension strategies to create awareness of the Internet, help local farmers acquire computer literacy, and persuade them to use the information and communication services. For Tongnan, Jinta and Pinggu, with the local governments acting as the change agencies, the extension of the Internet had become part of the local development agenda. The heavy involvement of rural administrations ensured the implementation as well as the supervision of the Internet projects.

In contrast with Internet diffusion in Yellow Sheep River, Jinta, Tongnan and Pinggu, the rural entrepreneurs at Shouguang and Huanglou adopted the Internet mainly under individual initiatives. The improved telecommunication infrastructures as well as the relatively developed rural economy created a favourable environment for the diffusion and adoption of advanced communication technologies.

With the overall social and economic development of Shouguang and Huanglou catching up with that of middle-sized cities, there was not much difference in terms of ways of Internet access between local rural dwellers and their urban counterparts, which could be easily achieved through the purchase of personal computers and Internet service subscriptions. However, despite farmers' increasing wealth, Shouguang and Huanglou did not witness wide adoption of the Internet among rural people. Instead, several entrepreneurs became the innovators in the rural communities by introducing and experimenting with the technology long before their peers. Due to the keen interest in technology, insatiable quest for scientific knowledge, and strong control of financial resources, the rural entrepreneurs pioneered in exploring the possible benefits of the Internet.

13.1.2 Internet Adoption

The two salient Internet diffusion models identified in this study provide rich sources to test the applicability of traditional innovation adoption concept in the rural settings. As Internet adoption under organizational intervention varies greatly from that under individual initiatives, the following discussion mainly aims at showing how the current cases in rural China may enrich our understanding of the concept.

While the diffusion of innovations theory focuses on the individual decision making process in explaining the possible adoption or rejection of an innovation, the findings from the Internet projects at Yellow Sheep River, Jinta, Tongnan and Pinggu imply that Internet adoption is not a matter of personal choice for those in the underdeveloped rural areas. As the weakest social class in terms of their share and control of social resources, rural people as individuals do not play an important role in the adoption of the Internet, which has to be the result of organizational initiatives at this stage.

When Internet diffusion is initiated by organizational forces, the decision making first takes place at the organizational level. One of the four types of innovation decisions proposed by Rogers (1983), the *contingent innovation decision*, seems to fit in with the case at hand. The contingent innovation decision holds that the adoption of an innovation is often determined by previous adoption decisions. Sometimes a prior innovation must be adopted at the organizational level before it becomes an option for individuals.

With regard to Internet adoption at Yellow Sheep River, Tongnan, Jinta and Pinggu, most of the decision making concerning *whether, when and how* to extend the Internet technology takes place at the *organizational* level rather than the *individual* level. It is either the investor company or the local governments that cover the cost of computer hardware and network facilities, subsidize Internet access, and invest in building local telecommunication infrastructures. With collective access in the form of information centres as the dominant Internet access mode, the costs associated with Internet adoption are evened out and shared among rural people who are likely to use and benefit from the technology.

Unlike individual Internet adoption in which the individuals are responsible for both the initial investment in computer hardware and the acquisition of technical know-how, the *hardware* and *"software"* parts of the innovation have been separately adopted by the change agencies and rural individuals. Internet adoption at the organizational level takes the form of group adoption with the collective purchase of hardware and improvement of infrastructures to facilitate the extension process.

However, Internet access itself does not accomplish the task of adoption. When the technology and services are made available to the targeted audiences, the decisions regarding whether to adopt the innovative way of getting information and communicating have to be made by individual farmers. It is the "software information"[25] imbedded in the Internet technology that the change agencies intend to extend among rural people.

In contrast with the adoption of old ICTs that makes less stringent demands on user skills, the adoption of the Internet requires a certain

25 Rogers (1983) argued that two kinds of information were normally involved regarding technological innovations: "software information" and "innovation-evaluation information" (p.14). "Software information" kept the potential adopters informed about the basic mechanisms and technical know-how about the innovations whereas the "innovation-evaluation information" dealt with the expected consequences of the technological innovations and their pros and cons so far as an individual adopter was concerned.

level of technical competence as well as some prior knowledge. Rural people have many difficulties in adopting the Internet since the majority of them have only acquired basic literacy and many do not know the Hanyu Pinyin Phonetic System and the methods of typing Chinese characters. Therefore, even though Internet access has been offered to farmers free of charge or at relatively low cost, the gap between the technology access and rural users is still huge. This explains why the Internet was compared to an aircraft carrier in the perception of farmers (according to the flower grower at Huanglou), because the two things are equally far away from their life.

Moreover, the Internet also differs from common agricultural innovations like fertilizers and new breeds of seeds, which are closely related to daily farming activities and may directly lead to increases in household earnings and cost savings. With high complexity and less relative advantages, compatibility, triability and observability, the Internet is therefore unlikely to be quickly accepted by rural people on a large scale. As a result, the extension efforts made by the change agencies mainly focus on creating awareness of the technology and encouraging rural people to utilize online services with the help of information technicians.

Therefore, access alone does not necessarily guarantee voluntary adoption in the rural settings. Internet adoption at the individual level is constrained by the general computer illiteracy and the comparatively low level of perceived efficacies of the Internet, especially when the technology is first introduced to the rural people. The findings across the four rural places under organizational intervention have revealed *two* dominant ways of Internet adoption as far as individual users are concerned – the adoption at the *awareness* level and the adoption at the *skill* level.

Internet adoption at the awareness level helps to describe the cognitive and attitudinal changes. The degree of awareness ranges from the knowledge of the existence of the technology, and the attitudes towards it, to a clear understanding of its information and communication functions through a mediated usage (such as reading web-based newspaper or information searching under the help of information technicians). With the E-paper delivered to over 90% of the county people, Jinta farmers realized the importance of online information to their farming decisions. During the three years when the MOST-UNDP project was going on, many rural people came to know the Internet through the bus banners, poster signs or handouts distributed. In Yellow Sheep River, farmers like Li Xingjiang and Wu Zhenzhou became aware of Internet services at the E-centre and personally benefited from the technology under manager Hu's help.

Internet adoption at the skill level involves the acquisition of E-literacy and first-hand experience in using the Internet, resulting in actual behavioural changes. Compared with the other three project sites in Internet adoption, Pinggu has the largest number of trained farmers. The individual capacity building is very likely to elicit voluntary use of the Internet and self-renewing efforts to continue the adoption behaviours.

Despite the fact that Shouguang and Huanglou are endowed with a favourable environment for diffusion of the Internet, the findings do not show impressive penetration of the Internet. In comparison with other places, the coastal rural sites suggest a different model of Internet adoption, which happens mainly as a result of individual choices.

At Shouguang and Huanglou, several rural entrepreneurs have started to invest in the technology. The characteristics of these people bear great resemblances to Rogers' (1983) description of innovators. With better financial resources, favourable attitudes towards technologies and a strong market orientation, the "vegetable doctor" Zhang, the grape master Huang and the flower grower Zhang adopted the Internet and utilized the technology at a time when the advantages of the technology and its cost-effectiveness were not so evident to the rural communities and even to themselves.

Shouguang and Huanglou have provided a unique case to study the natural diffusion process of the Internet in developed rural areas without external intervention. The anecdotal cases of Internet adoption suggest that financial affordability cannot be the sole factor that determines the adoption rate among farmers. The individual E-readiness, the perceived advantages of the Internet over the existing information and communication channels, and the cost-effectiveness of the investment are among the key factors that contribute to the reluctant attitudes held by local people towards the technology.

13.1.3 Internet Usage

As Internet adoption at the awareness level may not involve the actual usage of the technology, the concept of Internet usage is separately discussed to denote the employment of the Internet by rural users to meet concrete needs in daily life. This also serves as an attempt to overcome the pro-innovation bias so that more emphasis can be laid on the *post-adoption* stage of the innovation diffusion process.

When the Internet does not come to the rural people as a natural step in the social and economic development, the ways of deploying the technology are very much influenced by the change agencies. Since farmers depend on the financial and technical support provided by the governments or private companies to be introduced to the Internet, these organizations are obviously shaping their Internet usage through the extension initiatives. From the very beginning of their interaction with the technology, the Internet is perceived by farmers as a potential tool to improve living standards.

Business transactions via the E-centre in Yellow Sheep River made farmers like Wu Zhenzhou convinced that the Internet could be an efficient means to increase income, exchange information and establish contacts. The advertising campaign as well as the agricultural extension activities in Tongnan helped to frame farmers' perception about the Internet and attracted many of them to the UNDP centres for agricultural information and consultancy. The Information Office in Pinggu played an important role in leading Taoyuan farmers to employing the Internet to increase fruit sales. Therefore, rural users at the project sites took surfing online as a *purposive* activity from the very start. This is one of the important features that distinguish rural Internet users from their urban counterparts.

Another conspicuous feature of Internet usage at the project sites is the *mediated usage*. Except for Pinggu and Shouguang, farmers were rarely seen to surf the Internet by themselves. Most of the interaction between farmer users and the Internet was accomplished either through the mediation of people (information technicians or a few early adopters) or a different information channel (the E-paper in Jinta). This was largely due to rural users' low level of computer literacy and inability to process online information.

As a social phenomenon specific to the rural settings, the mediated usage is in effect a product moulded by the social contexts. On one hand, it helps to overcome the knowledge barriers. On the other hand, the mediated usage acts as an efficient information gatekeeping process, which facilities farmers' processing of web content and assists them in arriving at informed farming decisions.

Since agricultural productions are very much influenced by natural conditions like weather and soil, there are strong regional differences in terms of the applicability of farming skills and products. Rural users are confronted with high risks when making farming decisions on the basis of online information. The change agencies and information technicians play the important role of information gatekeepers to help farmers locate web content most relevant to local farming and reduce the uncertainties involved in adopting the information.

Take Tongnan as an example, when farmers came to the UNDP centres for solutions to farming problems, the information technicians often tried to meet their requests by sifting through the most relevant websites for useful and reliable information. If the results of the information searching were not satisfactory, the centres might turn to agricultural institutions for advice and additional help. Thereby the UNDP centres not only helped farmers ascertain the relevance and authenticity of web content but also managed to bridge the communication gaps between local agricultural institutions and their rural clients.

The majority of Jinta farmers also did not have to acquire computer literacy to use the Internet. The online information was disseminated in the form of a newspaper. This has greatly facilitated farmers' information processing as the print media were less skill-dependent. Most of the content of the E-paper was selected from agricultural websites both at the national level and the local level by the editors at Jinta Information and Network Centre. The content analysis of 114 issues of the *Herald* suggests that both the selection of the online information and the issuing of the newspaper were finely attuned to local farming cycles and farmers' information needs. Therefore, the *Herald* helped farmers arrive at sound production decisions by providing updated information on farming technologies, market trends, and rural policies.

The experiences of different rural settings have added variations to our conventional understanding of the concept of Internet usage, which usually implies direct operation of the technology. However, rural people have come up with a more pragmatic way of using the Internet. Through the mediation of information technicians or the traditional mass media farmers can still enjoy the benefits derived from the Internet without personally operating the computers.

When the Internet is introduced to the rural areas regardless of their readiness, the target users play an active role in the application of the technology. The individual usage driven by personal preferences very much determines the extent and ways that the Internet can be deployed to promote rural development. The findings of this study suggest that farmers' use of the Internet is closely associated with their farming or business activities. The behaviours of Internet usage are very much driven by information needs of the rural people, which motivate them to active information seeking either through direct online searching or mediated usage. In other words, farmers' usage of the Internet is driven by *information needs*, which are in turn determined by the *modes of production*.

Under the household contract responsibility system, the small scale self-subsistence economy becomes dominant in rural China. With each rural household as the basic economic and production unit, Chinese farmers are less likely to enjoy the production gains from scale of economies as their western counterparts. This is especially true for impoverished rural areas like Yellow Sheep River and Tongnan, where land cultivation was the sole means of subsistence. As the land output was just enough for the daily home consumption after the quotas fixed by the state (rural tax) being delivered, farmers did not have much disposable income to enlarge reproductions. Most of the cash transactions could be accomplished at the town markets where they sold sideline products in exchange for daily necessities.

Living in a traditional agricultural society of economic self sufficiency, farmers did not need much information on market trends or new farming practices since they had few marketable goods and their lack of resources made them less capable of bearing the possible losses involved in the trial of new seeds or farming technologies. Therefore, the information needs of rural people could be easily met by the existing communication channels like TV and word of mouth.

The backward production modes, low information needs and resistance to changes rendered the majority of local farmers at Yellow Sheep River or Tongnan reluctant to make use of Internet services even if they were aware of the tasks that the Internet could possibly accomplish. This helps to explain why farmers engaged in trading farm produces or the production of marketable goods were more enthusiastic about the information and communication services provided by the UNDP centres and benefited more from the Internet. The small volume of business transactions at the E-centre of Yellow Sheep River and the tiny number of farmers (2.2% of the county population) who used the Internet services in Tongnan demonstrate that local farmers did not take the Internet as an important tool for information and the technology seemed to bear little relevance to their development needs. The Internet did not help much in transforming the backward production modes and empowering the local people to get rid of poverty.

With the intensification of the rural economic reform and the gradual opening of the agricultural sector to international markets, better off farmers are able to use the opportunities to their advantage and make efforts to adapt to the transition from subsistence to market production. This has given rise to a more diversified rural economy, such as the cultivation of cash crops (cotton production in Jinta, and greenhouse farming and flower growing in Shouguang and Huanglou) and the involvement of farmers in trading agricultural products (fruit dealing in Taoyuan).

The information needs of farmers at these places have been greatly enhanced due to the transformation of the rural economic structures. Farmers have learned the hard way that increasing outputs will not necessarily result in rising income and therefore great efforts have been made to align household farming to market demands. Driven by a strong market orientation and a desire to get wealthy, farmers are increasingly aware of the important roles that information and knowledge play in modern agriculture. They actively seek information on new farming technologies and constantly keep track of price information and market fluctuations to reduce the level of risks in decision making.

Ideally, the Internet would serve as an efficient tool to enhance development for rural areas with a vibrant and prosperous economy. However, Jinta, Pinggu, Shouguang, and Huanglou provided contrasting experiences of using the Internet to meet the local development needs despite farmers' growing reliance on updated agricultural and market information to plan and organize daily productions. The empirical evidence suggests that the relevance of web content to the information needs of farmers is closely related to the perceived benefits of the Internet and greatly influenced by the effectiveness of the Internet in meeting these needs.

The popularity of the *Herald* among Jinta farmers was mainly due to the match between the selected web content and the information needs of local farmers. As more and more local farmers turned to greenhouse vegetable growing and livestock rearing, the *Herald* devoted much of its coverage to tips on the use of fertilizers and pesticides in greenhouses and various methods of diagnosing animal diseases and effective treatment. Therefore, the E-paper became an important source of information for local farmers to monitor new farming innovations and market changes.

In contrast, the Internet was not well received by farmers in Shouguang and Huanglou, due to its failure to offer satisfactory solutions for vegetable disease control or supply flower growers with updated knowledge on flower cultivation. Since most of the other rural places could not match Shouguang in terms of the development scale of greenhouse farming, they often look to Shouguang farmers for effective treatments of new diseases. This is the reason why "vegetable doctor" Zhang could not find much useful information over agricultural websites about the ways of controlling plant diseases in greenhouses. In Huanglou, likewise, the Internet could not compete with the existing information channels like farm newspapers or magazines in offering relevant content tailored to the information requests of flower growers. The shortage of specialized knowledge over the agricultural websites made local farmers turn to reading materials distributed at the annual flower expositions or popular journals on gardening.

13.1.4 Implications of the Internet on Rural Development

Despite so much hype around the impact of the Internet and the high hopes pinned upon it by the change agencies, the Internet has not brought about drastic social and economic advance to the project sites covered in this study. As far as the goal of the MOST-UNDP project was concerned, it failed to demonstrate its capacity in alleviating poverty though the diffusion of the Internet. At the planning stage of the project, it was hoped that the level of economic development for the towns and villages with Internet services would be five years ahead of those without. What disappointed the promoters was that the Internet did not make the project sites differ much from their neighbouring villages.

The initiator of the "Town and Talent Project", Mr. Wen Sayling, expected that Yellow Sheep River could leapfrog to an information society by spreading E-literacy and fostering knowledge-based economic development. He forecasted that the living standards of rural people at the western regions would be elevated to those of eastern coastal areas within ten years provided that the practices of Yellow Sheep River could be replicated in other impoverished areas. However, the remote town did not witness the fundamental changes in the life quality of rural people five years after the introduction of the Internet. The annual per capita income of local villagers slightly increased from RMB 1,060 (U.S.$133) in 2000 to RMB 1,123 (U.S.$140) in 2004.

Overall speaking, there are positive signs of social and economic progress as a result of the diffusion and usage of the Internet in the five rural places. However, the scope as well as the depth of the changes brought about by the technology are not significant enough to lend support to the romanticized arguments prevailing in the literature regarding the role of ICTs in general and the Internet in particular in promoting development. Internet diffusion and usage have not induced fundamental transformations in modes of production, organizations of work and consumption norms in the rural society.

Though the Internet did not bring the desired changes at the five rural places, the empirical findings of this study support the positive implications of the Internet on enhancing rural development. By referring to the human development concept, the following discussion intends to capture the implications that the Internet has on five dimensions of the life of rural users – livelihood, education, health, gender equality and political participation. As the main rationale behind the extension efforts of the change agencies is to use the Internet as an efficient tool to reduce poverty and improve rural economy, most of the repercussions have been felt at the economic sector of rural life.

The findings show that the Internet can contribute to improving agricultural productivity by increasing farmers' access to advanced agricultural technologies and facilitating information and knowledge exchange on farming innovations. The coverage of greenhouse cultivation technologies in the *Herald* helped many Jinta farmers learn to control the temperature and humidity in growing off-season vegetables. The E-paper also kept local farmers updated about the effective ways of controlling plant as well as livestock diseases, directly resulting in rising crop and livestock yields.

The market information and analysis on agricultural websites can assist farmers in making rational decisions and aligning farm output to market supplies and demands. For example, the UNDP centre became a reliable source of information for Liu Xiaolan to monitor the prices for various farm produces. By managing the sales according to price fluctuations at the local market, Liu was able to avoid the market glut and collect and sell grains at more profitable prices. The price information on the web enabled a few farmers at the neighbouring county of Yellow Sheep River to set a reasonable market price for the medical herbs they collected from mountains, whereas preciously they had to accept any prices offered by the middlemen.

The Internet has also been taken as an efficient communication channel by rural people to promote their farm produces and establish business contacts. In every rural place selected in this study, websites (for governments, villages, private companies, schools, or individuals) were set up to introduce the local economy, attract investment and enlarge business scales. On one hand, these websites served as windows through which rural people could look beyond their immediate milieus to seek business opportunities; on the other hand, they allowed these rural places to be known to the rest of the world and incorporated into the digital network.

Business transactions of over RMB1 billion were achieved in 12 towns of Jinta by posting information on the *Herald* and the local government websites in 2003. The village website of Taoyuan provided a business platform for local farmers to advertise fruit products and carry out transactions over the Internet. The online selling of fruits helped Taoyuan farmers avoid travelling great distances to seek potential trade partners and thereby greatly reduced the costs of sales. Within two years of individual adoption of the Internet, local fruit dealers managed to extend their businesses to many parts of the country and establish long term relationships with over 20 enterprises. The increasing orders propelled rural individuals to unite together so as to have organized sales of marketable surplus and bulk transport of fruit produces. The E-commerce created more business opportunities and led to a significant rise in average household earnings.

In terms of the possible implications of the Internet on rural education, the technology has been utilized either to enhance teaching quality and learning capabilities at rural schools or increase the level of E-readiness among adults. Whereas Yellow Sheep River achieved impressive results in using computers and the Internet to improve the middle school education, Pinggu made inspiring accomplishments in training farmers to acquire E-literacy.

The vocational middle school in Yellow Sheep River witnessed the inspiring changes brought about by the application of the Internet to the school education. The grasp of the computer operation skills enabled the teachers to be exposed to the vast amount of educational resources over the Internet. Through visiting the websites of elite middle schools in large cities, the teachers could improve the teaching methods without physically going to the places. Moreover, the active Internet usage was conducive to creating academic atmosphere and adopting modern ways of teaching and learning.

With about five to six students sharing one PC on average, the school enjoyed the best computer facilities in town. As young people were among the quickest in terms of reception of new technologies, the E-literacy acquired at an earlier age could have a long lasting impact on their future life. Even though the schoolers were only allowed to browse the school network, they were able to read the updated news and information about the outside, which might help to enlarge their horizon and gradually shape their perceptions of the world. The Internet stirred up a strong desire to change and the students anticipated a better life by acquiring knowledge and information. With the widespread usage of computer and network technologies on campus, the school underwent a steady increase of enrolment rate and the number of students admitted to high schools.

In addition to the positive impacts that the Internet has had on school teaching and learning, consistent efforts have been made by the change agencies to provide rural people with opportunities of learning computer and Internet application skills. Over 60 training sessions were organized by the E-centre in Yellow Sheep River to help entrepreneurial farmers, village chiefs and students grasp basic computer knowledge. The village UNDP centres in Tongnan conducted the trainings in several phases to target farmer groups with various educational levels. The most systematic and large-scaled E-literacy education was carried out at the information centre of Nandingfu Village in Pinggu. Together with 11 other information centres in the district, over 44,000 farmers (including information technicians) received formal training in four years, taking up almost 14% of the rural population.

Though the real changes that these one-shot trainings could bring to rural learners at the cognitive level remained doubtful and only a small percentage of farmers were able to operate computers and surf the Internet after the training, the promotion of E-literacy concomitant with the extension of the Internet has profound implications on narrowing the digital divide between urban and rural areas. The extension initiatives have enlarged chances for rural people to acquire knowledge on the Internet, the majority of whom would otherwise be totally deprived of the opportunities to know and appreciate the benefits brought by the advanced ICT.

The E-literacy training has also given rise to a number of information technicians, innovators and early adopters in rural areas. These rural talents could meet the severe shortage of technical personnel and therefore became the backbone of the rural communities in terms of providing technical support and maintenance services to governments, private organizations and individuals. Generally speaking, the Internet has been widely used by the change agencies to enhance capacity building, increase the level of individual E-readiness, and foster an environment conducive to the future employment of ICTs.

There has been little hard evidence that the technology is directly associated with the improvement of health, gender equality and political participation at the five rural places. This may be largely due to the fact that the Internet has been primarily used as a tool to improve agricultural productions and secure economic gains. However, several anecdotal incidents did shed some light on the contributions that the Internet could make to promoting healthy lifestyles, enhancing gender empowerment and providing social support, albeit in an indirect and subtle way.

The Internet services helped clinic owners in the neighbouring county of Yellow Sheep River and Tongnan get prevailing prices on medicine and establish contacts with pharmaceutical companies. Occasionally, farmers in Tongnan went to the UNDP centres to search possible treatment for uncommon diseases over the Internet. During the SARS outbreak in 2003, the *Herald* kept Jinta farmers informed of the typical symptoms of the epidemic as well as the precautionary measures. The E-paper also carried some online information on disease prevention and family planning from time to time.

In Yellow Sheep River, there was a general attitudinal change towards education among parents of female students after the Internet was introduced to the vocational middle school. The dropout rates of girls decreased and many chose to further their education or go outside to seek job opportunities after graduation rather than get married at a young age. This was definitely a symbol of social progress in the conservative mountainous town. In Machangying Town of Pinggu, female farmers accounted for 80% of the trainees of the E-literacy programmes. Wider access to education and employment opportunities enabled by the Internet might help to improve the living conditions of rural women and upgrade their social status.

Though the Internet had not been used for political purposes, two incidents in Yellow Sheep River suggested the Internet could empower rural people by lending them social support. The handicapped boy would not have won back his right if it had not been for information on relevant law cases the family searched over the Internet. The daughter of the teacher would not have received timely medical treatment without the social and financial support she obtained via the Internet.

13.1.5 The Issue of Sustainability

As the Internet involves considerable operating costs at the post-adoption stage, such as regular access fees and maintenance costs, this may to a great extent affect the sustainability of Internet services at rural areas. The Internet projects covered in this study have yielded valuable empirical data on running Internet services in rural areas with contrasting levels of socioeconomic development. Coincidentally, the Internet extension initiatives in Yellow Sheep River, Jinta, Tongnan, and Pinggu almost all commenced at the beginning of this century. The lapse of time enables us to compare the practical experiences among the rural places and identify the key factors that might affect the sustainability of Internet services.

As the only self-financed business model among the four places, the E-centre in Yellow Sheep River has been kept financially viable by charging tuition fees of the training sessions, daily Internet usage fees as well as commissions on a transaction basis. According to manager Hu, the total income of the E-centre was about RMB 3,000 (U.S.$375) in 2004, which was just enough for the expenditures on Internet access (RMB 200 per month, U.S.$25), electricity, phone calls, and hardware replacement. Hu did not expect the E-centre to make any money because neither the number of users of Internet services nor the amount of online transactions was likely to rise significantly.

Jinta and Nandingfu Village are two examples of Internet services funded and subsidized by the local government or rural collective. In Nandingfu, the Internet subscription fees (U.S.$500 per year) have been covered by the collective income of the village since the setting up of the information centre at the end of 2002. The large-scaled E-literacy training carried out at the information centre owes much to the strong village economy and the consistent efforts made by the village committee in improving basic infrastructure and social welfare.

Among the four rural places, Jinta has achieved the most satisfying results of Internet extension given the relative costs spent in extending the online information to over 90% of the rural households in the county. The adoption of the virtual server, the transmission of the *Herald* over the Internet, the employment of the existing network facilities and teachers at village schools, the selection of print media as the information carrier, and the recruitment of students in hand delivery have tremendously reduced the potential costs involved in extending online information to the vast rural mass. With the number of project villages reaching 75 by the end of 2004, the county government spent no more than U.S.$100 a year covering the costs of Internet access, printing paper and ink for each village.

Whereas the ongoing projects in Yellow Sheep River, Jinta and Nandingfu are still struggling for survival, the hype and excitement aroused by the Internet have already died out after the information centres at the project towns and villages in Tongnan closed down and many Taoyuan farmers sold their computers and gave up using the Internet in 2004. The previous analyses indicate that the low information needs of local farmers, the disempowerment of rural administrations, and the poor collective revenues foreclosed the grassroots initiatives in sustaining the Internet services in Tongnan. In Taoyuan, the intermittent Internet connectivity, the lack of timely technical support, and the shortage of collective income made local farmers lose interests in using the Internet.

13.2 Theoretical and Practical Contributions

13.2.1 Theoretical Contributions

The stagnation of development communication research has prompted this study to take a new perspective towards the role of communication technologies and communication in inducing development. Instead of viewing communication and development as measurable variables and aiming to establish a causal relationship, this study departs from the extant theoretical paradigms and takes a qualitative approach to seek an in-depth understanding about the possible linkages between the Internet and rural development. The experiences of the five rural places indicate that Internet diffusion and usage should be compatible with the existing political, socioeconomic, and technological contexts of the rural settings in order for the new ICT to serve as an engine for rural development. The interplay of structural factors and individual characteristics very much conditions the diffusion and usage of the Internet, which in turn determine the social and economic benefits accrued.

The specific contexts of rural China have provided an ideal testing ground to examine the applicability of the main concepts and propositions of the classical diffusion of innovations theory. As social observations should be interpreted and understood within a context (Whetten, 1989), the meaning of the Internet and its implications on rural development have been derived from the experiences of rural people in the five places characterized by varying socioeconomic development levels. In sharp contrast with the context of a well-developed market economy where the diffusion model originated on the basis of generalizing the behaviours of American farmers in the course of agricultural extensions, the specific characteristics of the selected rural places, people, and the innovation have generated new descriptions and explanations absent from previous studies.

First, the diffusion of innovations theory focuses on the individual decision making process in explaining the possible adoption or rejection of an innovation. Despite the fact that innovative ideas come from exogenous sources, the adoption decision is more often determined by the will of individuals. However, the findings of this study indicate that Internet adoption is not a matter of personal choice as far as the underdeveloped areas are concerned. As the most disadvantaged social group in terms of their share and control of social resources, rural residents as individuals may not play an important role in the adoption of the Internet.

In the context of rural China, where the local economy and infrastructure can hardly sustain such an advanced technology as the Internet, the diffusion and adoption of the Internet tend to be a result of organizational intervention. The change agencies play an essential role in extending the technology to the rural people, whose rationales for introducing the Internet greatly determine the ways of Internet access, the extension efforts taken to encourage farmers to adopt and use the technology, and the direction of social changes.

Under most circumstances, the innovation decisions first take place at the *organizational* level and then at the *individual* level. Whereas Internet adoption at the organizational level takes the form of group adoption with the collective purchase of hardware facilities and the establishment of information centres to provide public Internet access, the decisions regarding whether to adopt the innovative way of getting information and communicating are at the will of rural individuals. Therefore, the *hardware* and *"software"* parts of the Internet are separately adopted by the change agencies and rural individuals.

The separate adoption of hardware facilities by the change agencies and the functions of the Internet by rural users is a unique finding specific to the rural settings in this study. As many rural villages have not yet been reached by broadcast signals and telephones, it is virtually impractical to establish Internet access point for each rural household. The collective Internet access helps to distribute the hardware and connectivity cost across the large-sized rural population and therefore tremendously reduces the per capita cost of Internet access. With the initial investment in computer hardware shouldered collectively by the change agencies at the organizational level, the risks and uncertainties associated with individual Internet adoption have been very much lowered. The free or subsidized Internet access enables rural people to break the cost barrier and enlarges their chances of knowing the technology.

Subsequently, and as a second theoretical contribution, this study has added to the conventional understanding of the concepts of innovation adoption and usage in general and of Internet adoption and usage in particular. The experiences of the Internet projects indicate that ready access to the Internet does not necessarily guarantee voluntary adoption and active usage of the technology by rural individuals. The gap between access and adoption identified by this study reinforces the uncertainty in the implementation of acquired new media technologies as shown in the past research. It seems that the phenomenon of the assimilation gap not only exists in Western organizations and communities (Fichman & Kemerer, 1999; Steinfield, 2002; Wei, 2006), but also applies to rural users in China, the majority of whom are yet to acquire the intellectual and technical competence needed for processing online information. However, the finding unique to the rural settings in this study is that the five cases have manifested two variations of Internet adoption among rural people – adoption at the *awareness* stage and at the *skill* stage respectively – to differentiate *cognitive* and *attitudinal* changes from *behavioural* changes.

Internet adoption at the awareness level ranges from the knowledge of the existence of the technology, the attitudes developed, to a clear understanding about its various functions through a mediated usage. Most of the changes resulting from the extension efforts seem to have taken place at the cognitive and attitudinal level. Only a tiny fraction of the rural population at the project sites acquires E-literacy after training and thereby adopts the Internet at the skill level. The acquisition of Internet application skills is very likely to elicit voluntary use of the Internet and self-renewing efforts to continue the adoption behaviours.

As Internet adoption may not necessarily involve any interaction between rural people and the Internet at the behavioural level, the concept of Internet usage is introduced to denote the employment of the Internet by rural users to meet concrete needs in daily life. The findings of this study have demonstrated salient features that distinguish Internet usage in rural settings from that in urban settings.

Since farmers depend on the financial and technical support provided by the change agencies to be introduced to the Internet, these organizations are obviously shaping their Internet usage though the extension initiatives. The Internet is perceived by rural users as a potential tool to improve life quality and therefore online surfing becomes a *purposive* and *guided* behaviour from the very beginning.

Another unique feature specific to the rural settings is the *mediated* usage. Due to the general computer illiteracy and lack of capabilities to process online information, most of the Internet usage is accomplished either under the help of information technicians and early adopters or through reading online information carried by a print medium. The mediated usage overcomes the knowledge barrier posed by the Internet to rural users and facilitates their processing of online information.

Third, the current study has extended the existing knowledge about the influences of social structures on the diffusion process. The previous diffusion research has been criticized for the overemphasis on the impact of individual differences on adoption whereas the possible effects of motivations, perceptions and social contexts remain understudied (Zhu & He, 2002). This study has offered insightful understanding about how social structures exert influences on Internet diffusion, adoption and usage over and above individual characteristics. The findings of the study have provided strong empirical evidence in support of the social shaping of technology argument. The Internet serves as an agent of change and the extent, duration and intensity of its impact are primarily determined by the existing socioeconomic contexts of the rural settings.

Among the structural factors identified by this study, the organizational forces play a dominant role during the entire diffusion process. As the Internet does not come to the rural people as a natural stage of economic and infrastructural development in the rural areas, its diffusion and usage are strongly influenced by the change agencies, whose purposes to introduce the new technology shape the diffusion patterns of the Internet, which in turn condition the rural usage of the Internet and its implications for development.

However, despite the high hopes the change agencies have pinned upon the Internet to bring about intended changes, the degree to which the Internet can be adopted and used by rural people is conditioned by the joint influences of structural factors and individual differences. The individual Internet adoption and usage is greatly determined by the level of E-readiness, the information needs, and the perceived benefits of the Internet, the latter two of which are in turn determined by the production modes, the relevance of web content, and the existing information channels.

One of the unexpected findings of this study is that Internet adoption and usage tend to be closely associated with local farming and business activities. The information needs, as conditioned by the modes of production, motivate farmers to active information seeking, which contributes to adoption inclination and intensive usage. In other words, impoverished and conservative areas where self-sufficiency is the dominant economic mode may not have a heavy reliance on the Internet, whereas rural areas with relatively developed economy and a strong market orientation seem to be conducive to greater needs of information and potential wider Internet usage among farmers.

Subsequently, the availability of *relevant web content* to meet farmers' information needs generated in daily productions and the *relative advantage* of the Internet as compared to the conveniences provided by the competing media channels and content are decisive factors that affect Internet adoption and usage. Previous research findings on diffusion of new media technologies mainly indicate the importance of the information needs of potential adopters (Bowden, Kromer, & Tobia, 1994; Mullaly-Quijas, Ward, & Woelfl, 1994; Wallingford et al., 1996), but rarely touch upon the content that technologies carry. The findings of the study show that the degree to which the web content could satisfy the information needs may influence potential adopters' evaluation of the Internet, the decision-making process as well as usage of the technology at the post-adoption stage.

However, the effectiveness of the Internet in meeting the information needs is not solely determined by the relevance of the web content. Prior to the introduction of new ICTs, the rural communities rely on an existing communication network encompassing mainly old ICTs or interpersonal communication. Therefore, the availability and relevance of the web content, the cost-effectiveness of Internet adoption and usage, as well as the relative advantage of the Internet as compared to the existing media channels are among the key factors that shape Internet adoption and usage. These factors work together to motivate rural individuals either to experiment with the Internet and engage in self-renewing efforts to reinforce their behavioural changes or to discontinue the usage.

This study has also highlighted the key stakeholders involved in Internet diffusion and usage in the rural areas. The deployment of the Internet in promoting rural development depends very much on the perceived interests and benefits of rural administrators, telecommunication operators, private investors as well as rural individuals.

Besides making direct contributions to the diffusion of innovations theory, this study has provided rich empirical evidence regarding the actual impact of the Internet on the livelihood of rural people. Despite some positive implications that the Internet has on enhancing rural development, the diffusion of the new ICT has not resulted in drastic social and economic progress as expected. This study has presented a compelling counter argument against the leapfrogging literature and the optimistic views about the assumed roles of ICTs in poverty reduction and solving underdevelopment problems.

13.2.2 Practical Contributions

Though the Internet adoption is a relatively recent phenomenon in China, this study shows a variety of extension initiatives that have been developed in several rural places with contrasting socioeconomic and geographical conditions. The small scale and experimental applications of the Internet have demonstrated innovative and successful practices as well as pitfalls and failures. Despite the sporadic and isolated attempts, valuable lessons have been drawn from the explorative experiences to guide and direct future endeavours to sustain and replicate the Internet applications beyond the demonstration stage and broaden them in scope. As previously little is known about the use of the Internet by rural people in China, the cases examined by this study make meaningful comparisons with similar Internet application initiatives in other developing countries.

This study has led to several policy recommendations on the extension of the Internet and the provision of Internet services in the rural areas. First, it is recommended that governments, development planners and policy makers stop putting blind faith in technological changes. The Internet is unlikely to bring tangible benefits to rural people when there is no suitable environment to unleash its vast potentials.

The findings of this study show that in the short run, we cannot see a quick spread of Internet applications in the rural areas because the environment for a nationwide expansion has not matured yet and the information needs of the rural population have not evolved into a lucrative market force to sustain robust development. With the Internet as a relatively new tool of communication in China, the rural areas are certainly the least targeted sector for commercial Internet operators. As the costs of providing rural Internet services are relatively high and the demand for such services is comparatively low, there are fewer incentives for telecommunication operators to extend Internet services to the rural areas, especially those remote and mountainous regions.

Moreover, this study has not proven the effectiveness of the Internet in solving development problems. Overall, the project sites have not undergone significant improvement in social and economic conditions. This implies that the extension of the Internet should be compatible with rural development needs. In view of the high costs involved in its usage and the less appealing results, the Internet applications may not be taken as a top priority for impoverished areas. The cost-effectiveness of the Internet projects should be taken into serious consideration before any decisions on collective adoption should be made. More money should be spent in improving the basic physical infrastructures such as the construction of roads and rural schools, which may bear direct relevance to the livelihood of rural people.

Second, though the existing environments are not favourable towards the wide expansion of the Internet in the rural areas, experimental projects and isolated attempts should be encouraged to create awareness and provide rural people with opportunities of knowing the technology. To the vast rural population in China, the digital divide is not only a reality today but also seems to be an unbridgeable gap in both the near and distant future. It is unrealistic to think such a digital divide will be automatically reduced with the passage of time. As the Chinese economy is increasingly becoming knowledge-based, the disadvantages in communication and information exchange will certainly make the rural areas lag further behind rather than catch up with the urban areas.

At the current stage, the Internet application initiatives have to depend heavily on the strong commitment and support of local governments. The establishment of UNDP centres at Tongnan, Jinta Information and Network Centre, and the Information Office of Pinggu signifies the importance local governments have attached to the development of the Internet and its role in rural informatization. By incorporating Internet diffusion into the rural development strategies, the government administrations at various levels have mobilized the social, financial as well as human resources under their control to ensure the implementation of the Internet project as well as the achievement of the extension objectives.

Third, in addition to the heavy involvement of local governments, effective partnerships should be created among various stakeholders in the rural communities. One of the unique findings of the present study lies in the significant role that rural schools play in the extension of the Internet. The vocational middle school at Yellow Sheep River was selected by the change agency to train the local farmers and help them develop E-commerce. In Jinta, village schools also served as an important link to relay web-based content to the farmers. The MOST-UNDP project in Tongnan also situated its information centres at primary schools.

In these rural places, where the overall education level is inadequate to cope with an advanced communication technology as the Internet, schools as the powerhouse of knowledge in rural areas naturally help to fill in the gaps. As the central hubs of technological extension process, rural schools have not only increased teachers' and students' exposure to the Internet, but also facilitated the dissemination of advanced agricultural technologies and innovative farming and marketing ideas.

Last, the study identifies the feasible models of diffusing the Internet among the rural people. The collective Internet access proves to be one of the most cost-effective models given the large-sized population and weak infrastructural base of ICTs in the rural areas. The community Internet access points like the information centres can increase farmers' access to agricultural information and assist them in arriving at informed farming decisions. The information centres can also be used as training bases to elevate the technical competency of local people and contribute to the rural capacity building.

As charging for Internet services may curtail usage and participation, community information centres have to rely on the financial assistance provided by rural administrations like county or township governments and village committees to cover the running costs. However, there are still many ways of minimizing the costs to enhance future sustainability, such as purchasing second hand computers and working out contracts with Internet service providers to pay access fees on a collective basis rather than by the number of computers.

In addition to the establishment of collective access points, the multiple-step flow of online information is another effective way to help farmers cross the cost and knowledge barriers. The use of the print media to disseminate online information in Jinta provides one emulative model for providing information services in a cost-effective way. Jinta's experiences indicate that as the Internet usage involves a fair amount of technical know-how and the purchase of expensive computer hardware, online information is unlikely to flow to the farmers in one step. The additional steps in the flow of web content actually help to reduce the costs of Internet access, overcome the technical barriers, and facilitate the processing of information.

13.3 Conclusions

The findings of this study provide the following tentative answers to the three research questions posed by this study. With regard to the first research question, the results of this study show that the diffusion and usage of the Internet in the rural areas are greatly shaped by the political, socioeconomic, and technological contexts at the national level as well as those at the local level. Most of the extant social economic disparities between urban and rural people have been resulting from the household registration system and the rigid dual structure of urban versus rural areas. Rural people have been deprived of the opportunities of enjoying the social benefits provided by the state to urban residents, such as housing, employment, education and medical services. As a result of social closure, the majority of Chinese people have been excluded from sharing urban resources.

The population control system has become one of the root causes of such a huge divide between urban and rural areas in terms of various aspects of development. Despite the rapid growth of GDP over the last decade, the gap has been considerably widened rather than narrowed. Few countries have seen the inequality in income distribution as wide as in China in recent years. As a natural result of the socio-economic divide, the unbalanced needs for and access to information naturally follow.

Due to the underdeveloped economy and the inadequate tele-communication infrastructure, the rural western regions are lagging far behind the urban and coastal areas in respect of the diffusion and employment of both old and new media technologies. As a country so vast in territory and population, China is still at the stage of extending radio and television coverage to the remote areas. The universal access to electronic broadcast media has been taken by both national and local governments as a top priority in rural development strategies. Huge investments have been made to connect the last segment of 80 to 90 million rural people with broadcast signals and improve the overall qualities of rural broadcast services.

With regard to the diffusion of new ICTs, the Internet has undergone phenomenal growth in China since the late 1990s. However, Internet expansion is basically an urban phenomenon that excludes the majority living in rural and remote regions. The rural Internet users only account for about 1.4% of the total users (CNNIC, 2006). Though China's telecommunication industry has adopted the universal access obligation, the state-owned companies like China Telecom have only been urged to build infrastructures but not committed to providing services in the rural areas. Because the provision of Internet services among rural people is unlikely to generate profit, telecommunication operators are unwilling to extend their businesses. Therefore, unlike the extension of radio and TV under strong government intervention, the diffusion of the Internet or rather the lack of it in the rural areas has been driven very much by commercial interests.

With the existing economic and infrastructure environments of the rural areas unfavourable to the wide diffusion and usage of the Internet, quite a number of Internet application initiatives have been launched in recent years to test the possibility of using the Internet as a new engine for rural empowerment. Though the attempts are isolated and explorative in nature, the five rural places covered in this study have provided an enriched understanding on the social process of Internet diffusion and its implications on rural development.

The results of this study reveal that the diffusion and usage of the Internet have been conditioned by the interplay of structural factors and individual differences. For the majority of the rural places, the diffusion of the Internet tends to be a result of organizational intervention. The change agencies have played a dominant role in introducing the Internet to the rural people. The rationales behind the change agencies in introducing the Internet greatly determine the ways of Internet access, the extension efforts taken to encourage farmers to adopt and use the technology, and the direction of social changes.

Though the change agencies have promoted the Internet as a catalyst for desirable changes, the diffusion and usage of the Internet have been to a great extent conditioned by the rural economy. Due to the high costs involved in Internet usage, collective access points have been established at several project sites to facilitate farmers' adoption and usage of the Internet. The information centres provide farmers with information and communication services on a shared basis and therefore greatly improve the cost-effectiveness of Internet usage in these rural places. The public access to the Internet helps to remove the cost barrier posed by the technology to the rural users.

However, ready access to the Internet does not necessarily lead to voluntary adoption and active usage by rural people. Internet adoption and usage tend to be constrained by the level of computer literacy, the information needs, and the perceived efficacy of the Internet in meeting these needs as compared to the competing media channels and content. One of the most important findings of this study is that farmers' dependence on the Internet has been closely related to their production activities.

The impoverished rural places (for instance, Yellow Sheep River and Tongnan) where self-sufficiency remains the dominant economic mode seem less likely to rely on the Internet for information. The limited production activities do not require much information acquisition and what is available from TV and word of mouth already satisfies the local farmers. By contrast, in the relatively developed places (Jinta, Pinggu, Shouguang, and Huanglou) farmers' information needs have been greatly enhanced due to the transformations in rural economic structures. Motivated by a strong market orientation and a desire to get wealthy, farmers actively seek information on advanced agricultural technologies and market changes to reduce the level of uncertainties caused by information asymmetry and arrive at informed production decisions. Consequently, the Internet is more likely to be adopted and used by the active information seekers.

Subsequently, the relevance of web content and the cost effectiveness seem to be the decisive factors facilitating or impeding the adoption and usage of the Internet. Since rural people tend to be the least targeted consumers to both the service and content providers, the real benefits of the Internet for them are rather limited at the present stage. As a result, the Internet may give way to other popular information channels because of a lack of content with direct relevance to local farming.

In respect of the implications that the diffusion and usage of the Internet have upon various aspects of rural development, the findings of this study indicate that the degree to which the Internet can influence rural development rests with the ways the technology is diffused and used. The rural places covered in this study have not witnessed drastic social and economic changes resulting from the introduction of the Internet. Hence, the assumed roles of the Internet in enhancing rural development cannot be substantiated by this study.

Though the scope and depth of the changes are not significant enough to lend support to the romanticized arguments about ICTs for development, the empirical findings at the five rural places show some positive implications that the Internet has on various dimensions of the lives of rural people. As the Internet has been used by the change agencies as a tool to reduce poverty and improve rural economy, most of the satisfying results have taken place in the economic sector. The findings reveal that the Internet can help farmers improve agricultural productivity by increasing their access to advanced farming skills and facilitating information and knowledge exchange on farming innovations. The Internet has also proven to be an efficient communication channel for rural people to promote their farm produces and establish business contacts.

Besides improving the livelihood of rural people, many repercussions have been felt in the educational sector. Yellow Sheep River achieved impressive results in using networked computers to improve the middle school education. As the acquisition of computer literacy enabled the teachers to access vast education materials over the Internet, the wide use of the computers in lectures helped the students grasp information literacy and develop the habit of self-paced learning.

For the rural people at the project sites, the extension initiatives have raised their awareness about the Internet and provided them with opportunities of learning the technology. The Internet application initiatives have also given rise to a number of information technicians, innovators and early adopters in the rural places. Generally speaking, the E-literacy training has contributed to rural capacity building, increasing the level of E-readiness, and fostering an environment favourable to the future employment of ICTs.

This study has not produced many hard evidences that the Internet can be associated with the improvement of health, gender quality and political participation. This may be partly due to the reason that the Internet has been pushed to the rural people primarily as a tool to improve agricultural production and secure economic gains and the technology has not yet been widely used among the rural people.

In order to identify the roles of governments, private sectors, as well as individuals in harnessing the Internet for rural development and the practical ways of providing rural Internet services, this study has examined and discussed a variety of Internet application initiatives. Overall speaking, at the current level of rural economic and infrastructure development, the diffusion of the Internet has to rely mainly on planned efforts and strong support of local governments. As the rural areas tend to be the least targeted sector in terms of the development of the Internet, experimental Internet application initiatives should be encouraged to create awareness among rural people and provide them with opportunities of learning the technology. Otherwise rural people will be further marginalized at the periphery of the digital communication networks and denied the opportunities to use the Internet to benefit their lives.

This study has discussed many valuable experiences on running rural Internet services. The establishment of collective access points and the multiple-step flow of online information have proven to be cost-effective models of Internet usage. At the present stage, the public Internet services cannot rely on charging fees to cover the running costs since this may curtail usage and participation. Local governments should collaborate with the key stakeholders in the rural communities to come up with cost-effective ways of funding and subsidizing Internet services.

The explorative experiences of the five rural places indicate that the deployment of the Internet in promoting rural development depends much on the perceived interests and benefits of rural administrators, telecommunication operators, private donors as well as rural individuals. In order to provide sustainable rural Internet services, local governments need to create effective partnership with private organizations and telecommunication operators and solicit participation from the rural elites.

13.4 Limitations of the Study and Directions for Future Research

To date, with the overall economic and infrastructure environments of the rural areas unfavourable to the wide expansion of the Internet, the Internet application initiatives in the five rural places have been sporadic and isolated attempts. The limited diffusion does not allow me to foresee the possible social and economic impacts of the Internet when the extension of the technology in the rural areas reaches a full-blown stage. Consequently, the findings of this study as well as its conclusions may only apply to situations under the specific rural settings and may not predict other ongoing or future Internet applications across the country.

As the Internet adoption is a relatively recent phenomenon in China, most of the rural places covered in this study have started to experiment with the Internet since the beginning of the 21st century. My observations about the rural life and interactions with the rural people during the short trips only enabled me to capture the tangible benefits. Since it still takes time for many long term effects of the Internet to unfold, longitudinal studies should be carried out to provide for the understanding about the possible linkages between the Internet and rural development. More insightful and interesting findings would be generated if the same rural places could be revisited in five or ten years just as Lerner did in his study on the modernization process of the Turkish village.

Since all of the four pilot projects have been carried out for demonstration purposes, the artificiality of the environments may to a certain extent reduce the reliability of this study. Moreover, I relied on the support of the rural administrations to contact the local informants. It was highly likely that only those farmers who had had successful experiences of using the Internet were recommended, which gave me less opportunities of meeting up with those farmers who did not know about the Internet or who was aware about the technology but chose not to use it. Though participant observations and documentary analysis helped to verify and complement the data collected through in-depth interviews, more non-adopters should have been purposefully selected to understand their perceptions of the Internet. Therefore, future efforts should be devoted to examining the possible reasons behind the non-adoption of the Internet by rural people. This may lead to unexpected findings absent from the extant studies.

Bibliography

Academy for Educational Development. (1985). Beyond the flipchart: Three decades of development communication. Washington: Author.

Anderson, J., Crowder, L. V., & Dion, D. (1999). *Applying the lessons of participatory communication and training to rural telecenters.* Retrieved February 2, 2008, from <www.fao.org/WAICENT/FAOINFO/SUSTDEV/CDdirect/CDan0010.htm>.

Benjamin, P. (2001). The Gaseleka telecentre, northern province, South Africa. In C. Latchem & D. Walker (Eds.), *Perspectives on distance education telecentres: Case studies and key issues* (pp.75–84). Vancouver, Canada: The Commonwealth of Learning.

Best, M. L., & Maclay, C. M. (2002). *Community Internet access in rural areas: Solving the economic sustainability puzzle.* Retrievable up to November 17, 2004, from <www.cid.harvard.edu/cr/pdf/gitrr2002_ch08.pdf>.

Bhatnagar, S. & Vyas, N. (2001). *Gyandoot: Community-owned rural Internet kiosks.* Retrievable up to July 12, 2003, from <www1.worldbank.org/publicsector/egov/gyandootcs.htm>.

Bishop, R. L. (1989). Qi lai! Mobilizing one billion Chinese: The Chinese communication system. Ames, IA: Iowa State University Press.

Bowden, V. M., Kromer, M. E., & Tobia, R. C. (1994). Assessment of physicians' information needs in five Texas counties. *Bulletin of the Medical Library Association, 82*(2), 189–196.

Brewer, J. D. (2000). *Ethnography.* Buckingham, UK: Open University Press.

Brown, M. M. (2001). Can ICTs address the needs of the poor? *Choices (editorial)*. New York: The United Nations Development Programme.

Brynjolfsson, E., & Hitt, L. (1996). Paradox lost? Firm level evidence of the returns to information systems spending. *Management Science, 42*, 541–558.

Butcher, N. (1998). *The possibilities and pitfalls of harnessing ICTs to accelerate social development: A South African perspective.* Paper presented at the UNRISD Conference on Information Technologies and Social Development, Geneva.

Camacho, K. (2001). Evaluating the impact of the Internet in civil society organizations of Central America: A summary of a research framework. Retrievable up to November 19, 2004, from <www.acceso.or.cr/publica/telecom/frmwkENG.shtml>.

Carter, C. A., & Estrin, A. J. (2001, June). Market reforms versus structural reforms in rural China. *Journal of Comparative Economics, 29*(3), 527–541.

Chan, B. (2004). The Internet and the construction of Chinese cultural identity. Unpublished doctoral dissertation, Nanyang Technological University.

Chandler, D. (1995). *Technological or media determinism.* Retrieved February 9, 2008, from <www.aber.ac.uk/media/Documents/tecdet/tdet06.html>.

Chang, W. H. (1989). *Mass media in China: The history and the future.* Ames, IA: Iowa State University Press.

Chapel, R. (1980). Towards a new strategy for development. Oxford: Pergamon.

Chapman, R., & Slaymaker, T. (2002, November). *ICTs and rural development: Review of the literature, current interventions and opportunities for action.* Retrieved January 18, 2008, from <www.odi.org.uk/publications/working_papers/wp192.pdf>.

Chen, L. (1991). The door opens to a thousand blossoms: A preliminary study of communication and rural development. *Asian Journal of Communication, 1*(2), 103–121.

China Internet Network Information Centre. (1997, October). *Statistical report on the development of Chinese Internet*. Retrieved January 15, 2008, from <www.cnnic.cn/download/manual/en-reports/1.pdf>.

China Internet Network Information Centre. (2003). *The Internet timeline of China 1987-1996.* Retrieved January 15, 2008, from <www.cnnic.cn/html/Dir/2003/12/12/2000.htm>.

China Internet Network Information Centre. (2006, January). *The 17th statistical survey report on the Internet development in China.* Retrieved January 15, 2008, from <www.cnnic.cn/download/2006/17threport-en.pdf>.

China Journalism Yearbook 1982. (1982). Beijing: China's Social Sciences Press.

China Rural Technology Development Centre. (2005). *Jiaqiang nongcun xinxihua jianshe cujin pinggu jingji fazhan.* [Promoting the economy of Pinggu through building rural information network]. Retrievable up to January 1, 2006, from <www.crst.cn/infor/pinggu.htm>.

Chowdhury, N. (2000). *Information and communications technologies and IFPRI's Mandate: A conceptual framework.* Retrieved February 22, 2008, from <www.ifpri.org/divs/cd/dp/ictdp01.pdf>.

Chu, J. (1978). Broadcasting in the People's Republic of China. In J.A. Lent (Ed.), *Broadcasting in Asia and the Pacific* (pp. 36–37). Philadelphia: Temple University Press.

Chu, L. L. (1986). *Words into plowshares: Communication of agricultural innovations in Communist China.* Unpublished doctoral dissertation, Southern Illinois University at Carbondale.

Clarke, A. C. (2004). Foreword: Communications for goodness' sake. In UNDP (Ed.), *Promoting ICT for human development in Asia 2004: Realizing the Millennium Development Goals* (pp. iii–iv). New Delhi: Elsevier.

Constance, P. (1999). *1,800 teachers, one classroom.* Retrieved February 20, 2008, from <www.iadb.org/idbamerica/archive/stories/1999/eng/e299f8.htm>.

Crede, A., & Mansell, R. (Eds.). (1998). ICTs in developing countries (Booklet 1): The *importance for sustainable development.* The Hague: IICD.
Creswell, J. W. (1994). *Research design: Qualitative and quantitative approaches.* Thousand Oaks, California: Sage.
Cui, B., Lu, J., & Wang, X. (2005). Gaizhi yu zhuanzhe: 2004–2005nian zhongguo chuanmei chanye fazhan zongbaogao. [Structural reform and transition: Survey on the development of China's media industry, 2004–2005]. In B. Cui (Ed.), *Zhongguo chuanmei chanye fazhan baogao: 2004 zhi 2005 nian* (pp.3–33). [Report on development of China's media industry (2004–2005)]. Beijing, China: Social Sciences Academic Press.
Delman, J. (1993). Agricultural extension in Renshou County, China: A case-study of bureaucratic intervention for agricultural innovation and change. Hamburg: Institut fur Asienkunde.
Dos Santos, T. (1971). The structure of dependence. In K. T. Kan & D. C. Hodges (Eds.), *Readings in the U. S. imperialism* (pp.225–236). Boston: Extending Horizons.
Edelstein, A. (1984). Comparative communication research: A response to world problems. In B. Dervin & M. Voight (Eds.), *Progress in communication sciences* (pp.43–66). Norwood: Ablex.
Eisner, E. W. (1991). The enlightened eye: Qualitative inquiry and the enhancement of educational practice. New York: Macmillan.
Embassy of the People's Republic of China in India. (2005, February). *More sectors opened to private investors.* Retrieved January 17, 2008, from <www.chinaembassy.org.in/eng/zgbd/t184830.htm>.
Emmert, P. (1970). Complementarity in measurement. In P. Emmert & W. D. Brooks (Eds.), *Methods of research in communication* (pp. 161–164). Boston: Houghton Mifflin.
Federal Research Division of the Library of Congress. (2005). *China: Post-Mao Policies.* Retrieved January 7, 2008, from <www.country-data.com/cgi-bin/query/r-2745.html>.
Fei, X. (1989). *Rural development in China: Prospect and retrospect.* Chicago: The University of Chicago Press.

Fichman, R. G., & Kemerer, C. F. (1999, March). The illusory diffusion of innovation: An examination of assimilation gaps. *Information systems research, 10*(3), 255–275.

Frank, A. G. (1967). Capitalism and underdevelopment in Latin America. New York: Monthly Review Press.

Frank, A. G. (1969). *Latin America: Underdevelopment or revolution.* New York: Monthly Review Press.

Galtung, J. (1980). The true worlds: A transnational perspective. New York: Free Press.

Gansu Development and Reform Commission. (2005). *Gansusheng nongcun xinxi gonggong fuwu wangluo gongcheng shishi fang'an.* [The implementation plan of rural information service network in Gansu Province]. Gansu, China: Author.

Gao, F. (1991). The challenge of distance education in China. *The American Journal of Distance Education, 5*(2), 54–58.

Gilder, G. (2000). Telecosm: How infinite bandwidth will revolutionize our world. New York: Free Press.

Gold, R. L. (1958). Roles in sociological field observation. *Social Forces, 36,* 217–233.

Golding, P. (1974). Media role in national development: Critique of a theoretical orthodoxy. *Journal of Communication, 24*(3), 39–53.

Goonasekera, A. (1987). Third World cultures and a new information order. In N. Jayaweera & S. Amunugama (Eds.), *Rethinking development communication* (pp. 61–75). Singapore: AMIC.

Goonasekera, A. (1997). Asia and the information revolution: An introductory perspective. *Asian Journal of Communication, 7* (2), 12–33.

Grossman, L. K. (1999, July/August). *From Marconi to Murrow to – Drudge?* Retrieved January 23, 2008, from <backissues.cjrarchives.org/year/99/4/grossman.asp>.

Gulang gaikuang. (2005). [The overview of Gulang County]. Retrievable up to June 30, 2005, from <www.gulang.gansu.gov.cn/zoujin/view.asp?NewsID=247&classID=14>.

Hagen, E. E. (1962). *On the theory of social change: How economic growth begins.* Homewood, Illinois: The Dorsey Press, Inc.

Hammersley, M., & Atkinson, P. (1995). *Ethnography: Principles in practice* (2nd ed.). London: Routledge.

Hao, X., & Zhao, J. (2004a, May). *The wired farmers: Internet and social change in China's rural areas.* Paper presented at the 54th Annual Conference of the International Communication Association, New Orleans.

Hao, X., & Zhao, J. (2004b, July). *ICT for human development: Indicator or instigator?* Paper presented at the 13th Annual Conference of the Asian Media Information and Communication Centre (AMIC), Bangkok, Thailand.

Heeks, R. (1999, March). *The tyranny of participation in information systems: Learning from development projects.* Working paper no. 4 in Development Informatics series, University of Manchester.

Horton, D. E. (1991). Social scientists in international agricultural research: Ensuring relevance and contributing to the knowledge base. In W.F. Whyte (Ed.), *Participatory action research* (pp.218–236). Newbury Park, CA: Sage.

Hu, J. (2004, April). *China's development is an opportunity for Asia.* Speech presented at the opening ceremony of the Bo'ao Forum for Asia 2004 annual conference, Sanya, China. Retrievable up to January 12, 2006, from <www.chinaembassy-indonesia.or.id/eng/64392.html>.

Hu, W. (2005). [The E-commerce achieved by Yellow Sheep River vocational school]. Unpublished raw data.

Huang, C. (2003). Transitional media vs. normative theories: Schramm, Altschull, and China. *Journal of Communication, 53*(3), 444–459.

Huanglouzhen qingkuang. (2005). [An introduction to Huanglou Town]. Unpublished raw data.

Huiyuan xuexiao. (2005). [Member schools]. Retrieved January 30, 2008, from <www.yellowsheepriver.com/schoolbase/index.php>.

Information Office of the State Council. (2001, October 15). *Rural China's poverty reduction.* Retrieved February 20, 2008, from <english.people.com.cn/features/PRpaper/pr.html>.

Inkeles, A., & Smith, D. N. (1974). *Becoming modern: Individual change in six developing countries.* Massachusetts: Harvard University Press.

International Telecommunication Union. (2003). *World telecommunication development report 2003.* Retrieved January 12, 2008, from <www.itu.int/ITUD/ict/publications/wtdr_03/material/WTDR2003Sum_e.pdf>.

Internet World Stats. (2006). *World Internet users and population stats.* Retrievable up to November 16, 2006, from <www.internetworldstats.com/stats.htm>.

Jalava, J., & Pohjola, M. (2002). Economic growth in the new economy: Evidence from the advanced economies. *Information Economics and Policy, 14*(2), 189–210.

Jayaweera, N. (1987). Rethinking development communication: A holistic view. In N. Jayaweera & S. Amunugama (Eds.), *Rethinking development communication* (pp. 76–94). Singapore: AMIC.

Jiceng nianbao. (2005). [The annual report of Huanglou Town]. Unpublished raw data.

Jinta gaikuang. (2005). [The overview of Jinta County]. Retrievable up to October 23, 2005, from <www.jtxnet.com/jtgk/jtgk/jtgk.htm>.

Jorgenson, D. W. (2001, March). Information technology and the U. S. economy. *American Economic Review, 91*(1), 1–32.

Kenny, C. (2003). Development's false divide. *Foreign Policy, 4*(3), 76–78.

Krippendorff, S. (1979). The communication approach to development: A critical review. In J. A. Lent (Ed.), *Third World mass media: Issues, theory and research* (pp.71–82). Virginia: Williamsburg.

Lee, P. S. (1994, Summer). Mass communication and national development in China: Media roles reconsidered. *Journal of Communication, 44*(3), 22–37.

Lerner, D. (1958). The passing of traditional society: Modernizing the Middle East. US: Free press.

Letcher, P. (2004). *Country profile: China.* Retrieved January 25, 2008, from <www.itudaily.com/new/home.asp?articleid=4090802>.

Li, Q. (2004). *Nongmingong yu zhongguo shehui fenceng.* [Urban migrant workers and social stratification in China]. Beijing, China: Social Sciences Academic Press.

Lin, C-Y. (1993). The reform and development of China's sound broadcasting system in the reformation era (1976–1990). Unpublished master thesis, California State University.

Lincoln, Y. S., & Guba, E. G. (1985). *Naturalistic inquiry.* Beverly Hills, California: Sage.

Lindlof, T. R., & Taylor, B. C. (2002). *Qualitative communication research methods* (2nd ed.). Thousand Oaks, California: Sage.

Liu, A. P. L. (1964, June). *Radio broadcasting in Communist China* (Research Program on Problems of International Communication and Security). Cambridge, Massachusetts: Massachusetts Institute of Technology, Center for International Studies.

Liu, A. P. L. (1965, October). *The use of traditional media for modernization in Communist China* (Research Program on Problems of International Communication and Security). Cambridge, Massachusetts: Massachusetts Institute of Technology, Center for International Studies.

Liu, J., Meinzen-Dick, R., Qian, K., Zhang, L., & Jiang, L. (2005). *The impact of irrigation management transfer on household production in central China.* Retrieved January 10, 2008, from <www.cenet.org.cn/cn/CEAC/2005in/Renew35.doc>.

Long, G. (1995, October). The Internet as an electronic communications tool: Demystifying common utilities used to harness the Internet's strengths. *Telecom Asia, 6*(10), 42–47.

Long, J. (2001). *China's fiber boom: No end in sight.* Retrievable up to June 2, 2003, from <www.phoneplusinternational.com/articles/141sec5.html>.

Lu, X. (2002). *Sannonglun: Dangdai zhongguo nongye, nongcun, nongmin yanjiu.* [Research on the problems concerning agriculture, rural areas and farmers]. Beijing, China: Social Sciences Academic Press.

Lu, X. (Eds.). (2004). *Dangdai zhongguo shehui liudong.* [Social mobility in contemporary China]. Beijing: Social Sciences Documentation Publishing House.

Lull, J. (1991). China turned on: Television, reform, and resistance. New York: Routledge.

Mann, K. L., Eckert, S. E., & Knight, S. C. (2000). *A policy primer: Global electronic commerce.* Washington D.C.: Institute for International Economics.

Mansell, R., & Wehn, U. (Eds.). (1998). *Knowledge societies: Information technology for sustainable development.* New York: Oxford University Press Inc.

Marshall, C., & Rossman, G. B. (1989). *Designing qualitative research.* Newbury Park, California: Sage.

Massey, K., & Baran, S. J. (2001). *Introduction to telecommunications: Converging technologies.* California: Mayfield Publishing Company.

McAnany, E. G. (1980). The role of information in communicating with the rural poor: Some reflections. In E. G. McAnany (Ed.), *Communications in the rural Third World* (pp.3–18). New York: Praeger.

McClelland, D. C. (1961). *The achieving society.* Princeton, New Jersey: Van Nostrand.

McQuail, D. (1983). *Mass communication theory.* London: Sage.

Mead, M. (1970). Culture and commitment: A study of the generation gap. New York: Natural History Press.

Melkote, S. R., & Steeves, H. L. (2001).Communication for development in the third world: Theory and practice for empowerment (2nd ed.). New Delhi: Sage.

Merriam, S. B. (1988). Case study research in education: A qualitative approach. San Francisco: Jossey-Bass.

Miles, M. B., & Huberman, A. M. (1984). *Qualitative data analysis: A sourcebook of new methods.* Beverly Hills, California: Sage.
Ministry of Science and Technology. (2005a). *Zhongguo nongmin suzhi fazhan baogao 2004.* [The annual development report of farmers in China]. Beijing: Author.
Ministry of Science and Technology. (2005b). *Weifangshi ji dianxing-xian qingkuang.* [General information about Weifang City and its counties]. Retrieved up to December 6, 2005, from <www.crst.cn/infor/weifang.htm>.
Morse, J. M. (1991). Approaches to qualitative-quantitative methodological triangulation. *Nursing Research, 40*(1), 120–123.
Mowlana, H., & Wilson, L. J. (1990). The passing of modernity: Communication and the transformation of society. New York: Longman.
Mullaly-Quijas, P., Ward, D. H., & Woelfl, N. (1994). Using focus groups to discover health professionals' information needs: A regional marketing study. *Bulletin of the Medical Library Association, 82*(3), 305–311.
Murphy, R. (1990, September). Social closure: The theory of monopolization and exclusion. *The American Journal of Sociology, 96*(2), 480–482.
National Bureau of Statistics of China. (2005). *2004nian guomin jingji he shehui fazhan tongji gongbao.* [Statistics on 2004 national economy and social development]. Retrieved January 6, 2008, from <www.stats.gov.cn/tjgb/ndtjgb/qgndtjgb/t20050228_40223 1854.htm>.
Niu, Y. (2005). *Jinta nongcun jingji kuaisu zengzhang.* [Fast economic development in Jinta]. Retrieved February 10, 2008, from <gs.people.com.cn/GB/channel4/18/200502/09/22841.html>.
Obijiofor, L. M. (1994). Mass media and sociocultural phenomena in the process of development: An ethnographic study of two Nigerian communities. Unpublished doctoral dissertation, the Queensland University of Technology, Australia.
Okigbo, C. (1985). Is development communication a dead issue? *Media Development*, *4*, 23–25.

Oliner, S. D., & Sichel, D. E. (2000). The resurgence of growth in the late 1990s: Is information technology the story? *Journal of Economic Perspectives, 14*, 3–22.

Perroux, F. (1983). *A new concept of development.* Paris: UNESCO.

Pinggu Information Office. (2005). *Pinggu gaikuang.* [Pinggu overview]. Retrieved January 26, 2008, from <www.bjpgjw.gov.cn/>.

Qiang, C. Z.-W., Pitt, A., & Ayers, S. (2003). *Contribution of information communication technologies to growth.* World Bank Working Paper: 24. Washington, DC: World Bank.

Quibria, M. G., & Tschang, T. (2001, January). *Information and communication technology and poverty: An Asian perspective.* Retrieved January 22, 2008, from <www.globalknowledge.org/gkps_portal/aprm2002/workshop/ADBI.pdf>.

Richardson, D. (1998, June). The Internet and rural development. *The Journal of Development Communication, 9*(1), 54–67.

Rodan, G. (1998). The Internet and political control in Singapore. *Political Science Quarterly, 113*(1), 63–89.

Rogers, E. M. (1976, April). Communication and development: The passing of the dominant paradigm. *Communication Research, 3*(2), 213–240.

Rogers, E. M. (1983). *Diffusion of innovations* (3rd ed.). New York: The Free Press.

Rogers, E. M., & Scott, K. L. (1997). The diffusion of innovations model and outreach from the National Network of Libraries of Medicine to native American communities. Retrieved February 3, 2008, from <nnlm.gov/pnr/eval/rogers.html>.

Rogers, E. M., & Shoemaker, F. F. (1971). *Communication of innovations: A cross-cultural approach.* New York: Free Press.

Rogers, E. M., & Svenning, L. (1969). *Modernization among peasants: The impact of communication.* U. S.: Holt, Rinehart & Winston, Inc.

Röling, N. (1988). Extension science: Information systems in agricultural development. Cambridge, UK: Cambridge University Press.

Roling, N. G., Ascroft, J., & Chege, F. W. (1976). The diffusion of innovations and the issue of equality in rural development. In E. M. Rogers (Ed.), *Communication and development: Critical perspectives* (pp.63–78). London: Sage.

Rostow, W. W. (1971). *The stages of economic growth: A non-communist manifesto.* (2nd ed.). Cambridge: Cambridge University Press.

Ryan, B., & Gross, N. C. (1943). The diffusion of hybrid seed corn in two Iowa communities. *Rural Sociology, 8*, 15–24.

Sachs, W. (1992, June). Development: A guide to the ruins. *New Internationalist, 232,* 2–8.

Schramm, W. (1964). Mass media and national development: The role of information in the developing countries. California: Stanford University Press.

Schramm, W. (1976). End of an old paradigm? In W. Schramm & D. Lerner (Eds.), *Communication and change: The last ten years and the next* (pp. 45–48). Hawaii: The University Press of Hawaii Honolulu.

Schwandt, T. A. (1997). *Qualitative inquiry: A dictionary of terms.* Thousand Oaks, California: Sage.

Servaes, J. (1991). Toward a new perspective for communication and development. In F. L. Casmir (Ed.), *Communication in development* (pp. 51–85). New Jersey: Ablex Publishing Corporation.

Shen, J. (2004). *Zuihou yigongli: Jingji qianfada diqu jianshe nongye xinxi fuwu wangluo shiyong zhinan.* [The last mile connectivity: A handbook on establishing agriculture information services in less developed areas]. Lanzhou, China: Gansu People Publishing House.

Shingi, P. M., & Mody, B. (1976). The communication effects gap. In E. M. Rogers (Ed.), *Communication and development: Critical perspectives* (pp.79–98). London: Sage.

Shouguang gaimao. (2005). [Overview of Shouguang City]. Retrievable up to December 6, 2005, from <www.shouguang.gov.cn/sggm/default.asp>.

Shouguangshi gongzuo qingkuang huibao. (2005). [Shouguang government report]. Unpublished raw data.

Singhal, A., & Rogers, E. M. (1989). India's information revolution. New Delhi: Sage Publications India Pvt Ltd.

Singhal, A., & Sthapitanonda, P. (1996). The role of communication in development: Lessons learned from a critique of the dominant, dependency and alternative paradigms. *The Journal of Development Communication, 1*(7), 10–25.

So, A. Y. (1990). Social change and development: Modernization, dependency and world-system theories. California: Sage.

Sood, A. D. (2002). *Guide to ICTs for development.* Retrieved February 17, 2008, from <www.comminit.com/pdf/CKS_Guide_to_ICTs_for_Development.pdf>.

Sood, A. D. (2003). *Information nodes in the rural landscape.* Retrieved February 17, 2008, from <www.i4donline.net/issue/may03/pdfs/aditya.pdf>.

Spradley, J. P. (1979). *The ethnographic interview.* New York: Holt, Rinehart and Winston.

Steinfield, C. (2002). Organizations and external telecommunications networks: The adoption of wide area networks and their impact on organizational structure and strategy. In C. A. Lin & D. Atkin (Eds.), *Communication technology and society: Audience adoption and uses* (pp. 229–254). Cresskill, NJ: Hampton Press.

Stevenson, R. L. (1988). Communication, development, and the Third World. New York: Longman.

Tesch, R. (1990). Qualitative research: Analysis types and software tools. New York: Falmer.

Ulrich, P. (2003). China's rural Internet information centers: A project to reduce poverty through access to information and communication technologies (ICTs) in rural areas. Retrieved February 6, 2008, from <www.comminit.com/en/node/210226>.

Ulrich, P. (2004). Poverty reduction through access to information and communication technologies in rural areas: An analysis of the survey results from the social impact assessment conducted by the Chinese Ministry of Science & Technology and the United Nations Development Program. *The Electronic Journal on Information Systems in Developing Countries, 16*(7), 1–38.

United Nations Development Programme. (2001a). *Human development report 2001.* Retrieved February 5, 2008, from <hdr.undp.org/en/media/completenew1.pdf>.

United Nations Development Programme. (2001b). *Information communications technology for development.* Retrieved March 2, 2008, from <www.internetpolicy.net/practices/essentials5-web.pdf>.

United Nations Development Programme. (2003). *Human development report 2003.* Retrieved February 21, 2008, from <hdr.undp.org/en/media/hdr03_complete.pdf>.

United Nations Development Programme. (2004a). Promoting ICT for human development in Asia 2004: Realizing the Millennium Development Goals. New Delhi: Elsevier.

United Nations Development Programme. (2004b). *ICT and human development: Towards building a composite index for Asia.* Retrieved February 8, 2008, from <www.apdip.net/projects/rhdr/resources/idx_techppr.pdf>.

Van den Ban, A. W. & Hawkins, H. S. (1988). *Agricultural extension.* London, UK: Longman Scientific and Technical.

Wacker, G. (2000). *Behind the virtual wall.* Retrievable up to June 3, 2003, from <www.swp-berlin.org/biost/ber2000/b2000_06.htm>.

Wallingford, K. T., Ruffin, A. B., Ginter, K. A., Spann, M. L., Johnson, F. E., Dutcher, G. A., et al. (1996). Outreach activities of the National Library of Medicine: A five-year review. *Bulletin of the Medical Library Association, 84*(2), 1–60.

Wang, T. (2005). Ba cuncuntong zuowei guangbo dianshi wei sannong fuwu de zhongzhongzhizhong. [Setting “Extending Radio and TV Signals to Every Village” as top priorities among rural services]. *China Radio & TV Academic Journal, 170*, 6–9.

Wang, X. (2005). *Hulianwang yu qianfada diqu shehui fazhan yanjiu: Hulianwang zai xibu nongcun de liangzhong yingyong moshi de tantao.* [Research of the Internet and the social development in the developing regions: Discussion of two Internet application modes in the western rural China]. Unpublished doctoral dissertation, Communication University of China, China.

Wang, Z. (2005, July). *Issues on China agriculture management pattern and system.* Paper presented at the International Joint Conference for Commemoration of International Agreement on Academic Exchange between the Agricultural Economics Society of Japan and the Chinese Association of Agricultural Economists, Hokkaido, Japan. Retrieved January 10, 2008, from <www.soc.nii.ac.jp/aesj2/news/2005/aesj_caae3.pdf>.

Wang, G., & Dissanayake, W. (1984). Culture, development and change: Some explorative observations. In G. Wang & W. Dissanayake (Eds), *Continuity and change in communication systems* (pp.169–180). New Jersey: Ablex.

Weaver, J. H., & Jameson, K. (1978). *Economic development: Competing paradigms – Competing parables.* Washington, D. C.: Development Studies Program, Agency for International Development.

Weber, M. (1958). The Protestant ethic and the spirit of capitalism. New York: Scribner's Press.

Wei, R. (2006). Wi-Fi powered WLAN: When built, who will use it? Exploring predictors of wireless Internet adoption in the workplace. *Journal of Computer-Mediated Communication, 12*(1), article 5. Retrieved February 21, 2008, from <jcmc.indiana.edu/vol12/issue1/wei.html>.

Whetten, D. A. (1989). What constitutes a theoretical contribution? *Academy of Management Review, 14*(4), 490–495.

Whyte, W. F. (1991). Participatory strategies in agricultural research and development. In W. F. Whyte (Ed.), *Participatory action research* (pp.169–178). Newbury Park, California: Sage.

Wimmer, R. D. & Dominick, J. R. (2000). *Mass media research: An introduction* (6th ed.). Belmont, California: Wadsworth Publishing Company.

World Bank. (1995). *Increasing Internet connectivity in Sub-Saharan Africa: Issues, options, and World Bank group role.* Retrieved January 23, 2008, from <www.uneca.org/eca_resources/Major_ECA_Websites/padis/telemat/africa03.htm>.

World Bank. (2001). *Global economic prospects and the developing countries 2002.* Retrieved January 24, 2008, from < web.worldbank.org/WBSITE/EXTERNAL/EXTDEC/EXTDECPROSPECTS/GEPEXT/EXTGEP2002/0,,menuPK:544363~pagePK:64167702~piPK:64167676~theSitePK:544346,00.html>.

World Economic Forum. (2003). The global information technology report 2003–*2004.* Retrieved February 21, 2008, from <www.weforum.org/pdf/Gcr/GITR_2003_2004/Executive_Summary.pdf>.

Ye, J. (2002). Processes of enlightenment: Farmer initiatives in rural development in China. Unpublished doctoral dissertation, Wageningen University. Retrieved January 10, 2008, from <library.wur.nl/wda/dissertations/dis3281.pdf>.

Yu, F. T. C. (1977). Communication policy and planning for development: Some notes on research. In D. Lerner & L. M. Nelson (Eds.), *Communication research: A half century appraisal* (pp. 167–190). Honolulu: University Press of Hawaii.

Yu, Y. (2005, April). *The "Extending Radio and TV Signals to Every Village" project.* Speech presented at the Municipal Radio and TV conference, Chongqing, China. Retrieved January 12, 2008, from <www.cq.gov.cn/zwgk/ldjh/20050510164.HTM>.

Zhang, H. (2005, May). Anzhao kexue fazhanguan de yaoqiu quanmian jiaqiang nongcun guangbo dianshi gongzuo. [Improving rural electronic broadcast according to the concept of overall development]. *China Radio & TV Academic Journal, 170*, 9–11.

Zhang, Z. (2005, May). Qianghua fuwu "sannong" shi meiti de lishi zeren. [Serving rural people is the historical responsibility of the mass media]. *China Radio & TV Academic Journal, 170*, 12–14.

Zhang, Z. (2005, May). Qianghua fuwu "sannong" shi meiti de lishi zeren. [Serving rural people is the historical responsibility of the mass media]. *China Radio & TV Academic Journal, 170*, 12–14.

Zhongguo zheng zhuoshou zhiding xinxi chanye wudaoshinian de zhengce gangyao. [China is making development plans for its information industry]. (2004). Retrievable up to December 26, 2005, from <51sobu.com/dongtai/content/2004413281083130913734.html>.

Zhu, Q. (2002). *Nongcun shehuixue*. [Rural sociology]. Beijing, China: China Agriculture Publishing House.

Zhu, J., & He, Z. (2002). Perceived characteristics, perceived needs, and perceived popularity: Adoption and use of the Internet in China. *Communication Research, 29*(4), 466–495.

www.ingramcontent.com/pod-product-compliance
Ingram Content Group UK Ltd.
Pitfield, Milton Keynes, MK11 3LW, UK
UKHW021843100926
8556IPUK00019B/1013

9 783039 115846